The UK Environmental Foresight Project

VOLUME 2

Road Transport and the Environment: The Future Agenda in the UK

Air Pollution

KEITH D MASON

London HMSO

© Crown copyright 1993
Applications for reproduction should be made to HMSO
First published 1993

ISBN 0 11 752841 2

For further information about this publication contact
Mr Keith Mason or Dr Robert Whelan at:

CEST
5 Berners Road
Islington
London N1 0PW

Tel: 071-354 9942
Fax: 071-354 4301

Cover Photo: Tom Van Sant
Science Photo Library

Forward

In its 1993 White Paper on Science and Technology, the Government made clear the importance that it attaches to the process of technology foresight in drawing up the nation's science and technology agenda. Foresight must also play a vital role in developing the nation's environmental technology and policy. Cleaning up environmental problems after they occur is invariably more expensive than acting ahead to avoid them in the first place. Foresight can help in preventing pollution as well as in preparing an appropriate response.

We have been happy to initiate the UK Environmental Foresight Project with the Centre for Exploitation of Science and Technology and assist with its funding. We are especially grateful for those individuals and organisations who took part in the process. We will be studying the findings carefully. We are already putting into use some of the lessons learnt from the consultative process. We hope that they will also be valuable to the drawing up of the national Technology Foresight programme announced earlier this year.

Dr D J Fisk
Chief Scientist
Department of Environment

Environmental foresight is a vital component of strategic industrial management. In order to focus investments in innovation, the future issues of environmental concern must be contemplated today. The environmental agenda of the future will require innovative technology and creative public policy. Solutions to future environmental issues must enable firms to operate 'green', while remaining in the 'black'. New technology, as well as an expanded application of existing technology, must be a part of those solutions. The Environmental Foresight Project provides a useful first step in looking ahead and outward.

The transport sector of the economy provides a key link between the UK and the rest of the world. Environmental demands from abroad are changing the technical rules of the game here at home. UK industry has the potential to respond to the future transport challenges with new, more environmentally benign, transport fuels and the next generation of motor vehicle technology itself. By illuminating the goals of a strategic transport and environment policy, this volume of the Foresight Project will help position UK industry and government for the future.

CEST would like to thank the Department of Environment for its support and sponsorship. We hope that the specific lessons learned, as well as the methodology employed, will provide a valuable contribution to the broader foresight deliberations that are unfolding here in the UK. We hope the project will serve to stimulate debate, enhance collaboration, and prepare the UK for the environmental challenges of the future.

Dr R C Whelan
Chief Executive
CEST

Preface

About the UK Environmental Foresight Project

The UK Department of Environment and CEST initiated the Environmental Foresight Project in August 1992. The project is designed to help the UK government, industry and researchers prepare for the environmental agenda of the future.

This volume assesses an important part of the future transport-related environmental agenda for the UK – air pollution. This volume provides a comparative analysis of road transport air pollution policies in other countries and discusses their relevance for the UK situation.

The companion Environmental Foresight Project Volume 1 (published separately) discusses the National Foresight Workshop and explores other countries' environmental planning and foresight activities. In addition, Volume 1 explores the relationship between UK environmental foresight and the global sustainable development agenda. Volume 3 of the project continues this volume's detailed exploration of road transport and the environment by assessing the future road transport noise agenda.

About the Centre for Exploitation of Science and Technology

CEST identifies emerging issues which could give rise to significant commercial and technological opportunities for industry. Working with companies, scientists and government policy makers, CEST identifies opportunities and stimulates companies to act on them. Projects are participatory and designed to help companies perceive new opportunities earlier and hence reduce the leadtime for companies to gain commercial benefit. CEST's perspective is global, with particular focus on Europe, USA and Japan. Founded by industry in 1988, CEST is funded by a consortium of over 30 companies and government departments. CEST is independent, has strong links with government and higher education and its work is public and widely disseminated.

About the Author

Keith Mason joined CEST in 1992 on leave from the US Environmental Protection Agency in Washington, DC where he was the Senior Economic Analyst for the new US Clean Air Act of 1990.

Acknowledgements

The Environmental Foresight Project has benefitted from the guidance and participation of many individuals involved in the National Foresight Workshop (discussed in Volume 1) and those assisting with the present volume. Special thanks is extended to Dr Peter Saunders of the Chief Scientist Group of the

UK Department of Environment (DoE) and other members of DoE for their guidance and to my colleagues at CEST for their important contributions to the project.

Special thanks to Rebecca Nancarrow for manuscript preparation and project assistance.

Please note that the views expressed in this report are not necessarily those of the Department of the Environment, the Council of CEST or of any individual or organisation involved in the Phase I National Workshop.

Executive Summary

Only by considering all dynamic aspects of the UK economy and society will successful preparation for the environmental agenda of the future occur. Transport will play a key role in defining the environmental challenges facing the UK after the turn of the century. The virtual explosion in the number of vehicles on the roads in the UK, and the distances those vehicles travel, pose a vital environmental challenge that requires strategic consideration by government, industry and the public.

This report describes how road transport contributes to one of the most pervasive environmental problems on today's agenda – air pollution. By outlining the changing nature of road transport air pollution in the UK, this report will illustrate a possible future agenda and likely policy and technical responses.

Policies aimed at motor vehicle pollution are unique. Unlike most other types of environmental policies, they have a unique quality to seek near universal adaptation – albeit at different times and in different forms according to the geographic markets considered. To improve the UK's preparation, it is vital to look at the past trends, current positions and future directions of vehicle emission control policies around the world.

By noting the direction that transport environmental policy and technology may develop, and by improving the understanding of air pollution within the UK, the future challenge can be outlined and the strategic capacity to respond enhanced. Without an integrated strategic assessment of future transport policy, opportunities for an improved quality of life and economy will simply not be realized.

Road Transport Environmental Policy Trends

The amount of air pollution generated by road transport in the UK is directly related to emission control policies of all types. The future UK policies addressing motor vehicle pollution will be strongly influenced by policy developments abroad. These policies are part of clearly discernible trends which can be projected into the future.

Concerning air pollution from passenger cars, major trends discussed in this volume include:

A Changing Policy Goal

Emission control policies originally aimed at conventional air pollutants (eg NO_x, HC, CO) have expanded to address global pollutants (eg CO_2, CFCs) and hazardous air pollutants (eg benzene, 1,3-butadiene).

A Changing Scope of Motor Vehicle Policy

Motor vehicle air pollution was originally considered a local problem. This scope of concern has expanded to a regional context with the improvement of air chemistry knowledge and models. It now encompasses a global scope. The next scope for policy development is likely to be at a local 'micro' environmental level due to increased concerns for elevated levels of air pollutants found in certain urban situations such as near roads and within vehicles themselves.

A Move towards Local Air Quality Management

Due to inadequate regional or national management of air quality, local air quality management areas have developed with planning responsibilities and the capability to implement local control measures. This trend will continue and should be matched by improvements in local jurisdiction.

Expanding Focus of Emission Controls

Vehicle emission control programmes have expanded from the exhaust tailpipe to other sources of vehicle pollution not addressed previously. 'In-use' emissions are being regulated through vehicle inspection and maintenance programs and "evaporative" emissions are being controlled through modified vehicle testing and certification procedures.

From Vehicles to Fuels to Fuel Distribution

In addition to the expanding control of emissions emanating from the vehicles themselves, vehicle fuels are becoming controlled. Both fuel quality and mandates for alternative non-crude transport fuels are being considered and implemented.

From California to Europe

Motor vehicle emission control policies for the past 25 years have been initiated in California first, next adopted by the US, and then by the EC, Japan and Scandinavia. This trend, for the most part, should continue for the next decade. Traffic planning measures are more likely to originate in Europe first, rather than California, however. Europe also has a stronger lead in integrating energy and environmental policy which, in the case of transport, leads to developments such as the life-cycle energy analysis of vehicle fuels and the use of fuel taxes to limit demand.

From Cars to Trucks to Fleets

Traditionally, passenger cars have drawn the most attention from policy makers. As cars become relatively more controlled, attention will switch to less controlled sources, such as lorries, which will make up a relatively greater proportion of road transport emissions. As alternative fuels become a subject for policy consideration and as air quality management switches to local jurisdictions, policies addressing fleets of vehicles will become commonplace.

From Vehicle Performance to Vehicle Use

As the technical options for vehicle emission control are implemented, attention will turn to the emission performance of vehicles under different driving conditions. The most likely type of vehicle use policy will affect vehicle speed.

From Vehicle Regulation to Transport Planning

Eventually, the growth in either the number of vehicles or the miles that they drive will result in the necessary application of transport planning and various restrictions on either the demand for private transport or the supply of it. Unfortunately, transport demand management and planning have been applied in most circumstances as an adjunct to vehicle-based emission control policies. In the future they will need to play an equally important role.

From 'Command and Control' Regulations to Market Forces

Traditionally, road transport air pollution has been addressed through specific regulations on vehicle emissions and fuel quality. New environmental issues associated with motor vehicles, such as CO_2 emissions and global warming, lend themselves to more market-based policy methods. Increased fuel taxes and road pricing to reduce demand for passenger car transport are being implemented or experimented with in many countries, including the UK. Emissions trading policies, similar to those enacted to address acid rain emissions in the US, are being contemplated for mobile sources.

From Mobile to Stationary Sources

As mobile sources are progressively more controlled, and as air pollution problems traditionally associated with mobile sources persist, stationary sources of air pollution will come under control. This integration of mobile and stationary source control measures will allow for a more efficient set of control policies.

By noting and evaluating these and other trends, the UK will be in a clear position to evaluate the necessity for such new policy developments given its own pollution problems. Some of these developments, necessary or not, will affect the UK market and policy arena. By understanding the direction of world transport environmental policy, the UK will be able to prepare for the implementation and technical requirements that these trends represent.

Contents

1	**Introduction**	1
2	**The Motor Vehicle Contribution to Air Pollution in the UK**	5
	2.1 UK Road Transport Air Pollution Trends in Comparison	7
	2.2 The Significance of Road Transport Emissions in Urban Environments	11
	2.3 Vehicle Type and Road Transport Emissions	11
	2.4 The Contribution of Fuels to UK Road Transport Air Pollution Emissions	15
	2.5 Road Transport Contribution to UK VOC Emission Inventories	15
	2.6 Road Transport Contribution to UK CO_2 Emissions	17
3	**UK Ambient Air Quality Concentrations and Trends to Date**	19
	3.1 Nitrogen Oxides Concentrations and Trends to Date	19
	3.2 Tropospheric Ground-Level Ozone Concentrations and Trends to Date	20
	3.3 Particulate Matter Concentrations and Trends to Date	20
	3.4 Carbon Monoxide Concentrations and Trends	21
	3.5 Lead Ambient Concentrations and Trends	21
	3.6 Future UK Road Transport Emission Projections	22
4	**UK Air Pollutant Compliance with Standards, Guidelines and Directives**	27
	4.1 Carbon Monoxide Levels and Compliance	27
	4.2 Nitrogen Dioxide Levels and Compliance	29
	4.3 Particulate Matter Levels and Compliance	31
	4.4 Ozone Levels and Compliance	33
	4.5 CO_2 Emission Levels and Progress towards Targets	35
5	**Global Air Pollution and Road Transport**	37
	5.1 Global Warming, CO_2 Emissions and Transport	37
	5.2 Stratospheric Ozone Depletion and Road Transport	45
6	**The Health Effects of Road Transport Air Pollution: Current and Future Policy Developments**	53
	6.1 Introduction	53
	6.2 Developments in 'Conventional' Air Pollution Health Research	54

6.3	Mobile Source Hazardous Air Pollutants	57
6.4	Human Health Risk Estimates Due to Toxic Air Pollutant Emissions	65
6.5	Other Mobile Source Air Toxic Information and Studies	70
6.6	Air Toxic Exposure Situations	73

7 The Regulation of Motor Vehicle Air Pollution — 75

7.1	A Short History	75
7.2	Notable Trends In US and World Passenger Car Air Pollution Control	76
7.3	Global Warming: The Merging of Fuel Economy and Environmental Policy	82
7.4	Recent Developments in Automobile Emission Control Policy	83
7.5	Passenger Car Tailpipe Emission Control Measures	102
7.6	Vehicle Speed and Emission Control	110

8 Motor Vehicle Fuel Quality Regulations — 115

8.1	Lead Content in Petrol	115
8.2	Benzene Limits in Petrol	117
8.3	Reformulated 'Cleaner' Gasoline	119
8.4	Diesel Fuel Quality	122
8.5	Oxygenated Fuels	128
8.6	Alternative Fuels	131
8.7	Ethanol and Biofuels: Agricultural Policy, Energy Security and Global Warming	133

9 Conclusions — 137

9.1	Foresight and Transport	137
9.2	Road Transport and Air Quality	138
9.3	Road Transport Policy Trends	139
9.4	Potential UK Policy Developments	141

Tables

Table 1	Sources of the Principal Pollutants In the UK (1990)	5
Table 2	Contribution of Sources to Air Pollution in the EC (%) 1987	6
Table 3	Road Transport Contribution to Air Pollution in the UK	6
Table 4	Road Transport Contribution to Air Pollution in the US	7
Table 5	Percentage Changes in UK and US Road Transport Vehicle Emissions	8

Table 6	Passenger car contribution to total UK road transport emissions (1990)	13
Table 7	Freight (HGV) Percentage Contribution to Total UK Transport Air Pollution	13
Table 8	Fuel Contributions to UK Emissions	15
Table 9	Past trends and projections of UK CO_2 emissions by sector (MtC)	17
Table 10	CO Ambient Standards and Guidelines	27
Table 11	NO_2 Ambient Air Quality Standards	29
Table 12	NO_2 National Air Quality Public Health Indices	29
Table 13	Particulate Matter Ambient Air Quality Standards	31
Table 14	Particulate Matter Public Health Indices	31
Table 15	National Air Quality Public Health Indices O_3	33
Table 16	UK Passenger Car Greenhouse Gas Emissions	38
Table 17	UK Consumption of CFCs and Halons	47
Table 18	UK Consumption of CFCs by Primary Area of Application	47
Table 19	Stratospheric Ozone Depleter Phaseout Deadlines	51
Table 20	Motor Vehicle Contribution to Ambient Air Toxic Concentrations	58
Table 21	US Mobile Source Cancer Study: Emission Factors and Incidence	66–67
Table 22	In-Use Passenger Car Inspection and Maintenance Tests and Programmes	87
Table 23	California Off-Road Diesel Vehicle Emission Standards	101
Table 24	HC & NO_x Combined Passenger Car Emission Standards	103
Table 25	Carbon Monoxide Passenger Car Emission Standards	104
Table 26	HC Passenger Car Emission Standards	104
Table 27	NO_x Passenger Car Emission Standards	105
Table 28	EC Passenger Car Emission Standards	105
Table 29	Emissions Benefits from Controlling Car Speeds	111
Table 30	Current Road Use and Car Speed Characteristics	112
Table 31	Market Share of Unleaded Petrol	116
Table 32	Benzene in Petrol Content Limits	118
Table 33	US Reformulated Gasoline Specifications	120
Table 34	California "Phase II" Reformulated Gasoline Specifications	120
Table 35	Assumed penetration of diesel cars in the UK new car market	123
Table 36	Diesel Fuel Sulphur Content Limits	124
Table 37	Sweden 1992 Diesel Fuel Classifications	125
Table 38	Diesel Fuel Introduction in Buses in Denmark	126
Table 39	California Diesel Reference Fuel Specification	126
Table 40	Clean Diesel Tax Incentive Policies	127

Table 41	EC Oxygenates limits	129
Table 42	15% MTBE and 10% Ethanol Emission Adjustment	131
Table 43	Percent Change in Air Toxics Levels for M85 and M100 Relative to Gasoline	133

Figures

Figure 1	Project Scope	2
Figure 2	Road Transport Contribution to UK & US NO_x Emissions	8
Figure 3	Road Transport Contribution to UK & US Carbon Monoxide Emissions	9
Figure 4	Road Transport Contribution to UK and US HC Emissions	9
Figure 5	Road Transport Contribution to UK & US Particulate Matter Emissions	10
Figure 6	Sources of SO_2 in Greater London	11
Figure 7	Sources of NO_x in Greater London	12
Figure 8	Sources of Black Smoke in Greater London	12
Figure 9	NO_x emissions from Road Transport in the UK	14
Figure 10	UK Road Transport Emissions of NO_x	16
Figure 11	UK Road Transport Emissions of CO_2	16
Figure 12	Carbon dioxide emissions by end user, 1990, UK	18
Figure 13	Mean Urban Concentration of Black Smoke in the UK	21
Figure 14	Annual Mean Concentrations (ng/m^{-3}) of Lead in UK Urban Air	22
Figure 15	UK Passenger Car VOC Emission Projections 1990–2025	23
Figure 16	UK Passenger Car NO_x Emission Projections 1990–2025	23
Figure 17	UK Passenger Car CO Emission Projections 1990–2025	24
Figure 18	UK Passenger Car CO_2 Emission Projections 1990–2010	25
Figure 19	Hourly ozone concentrations classed as "poor" at selected UK sites	35
Figure 20	Ozone levels in the Antarctic, 1957 to 1990	46
Figure 21	Relative Contribution by Source Categories to Total Estimated Cancer Cases per Year in the USA	59
Figure 22	HC emissions from average model – year car (g/mile)	95
Figure 23	Vehicle Age and Mileage	96

1 Introduction

The results of the first phase of the Environmental Foresight Project indicate two areas of the UK economy that will be at the heart of a future UK environmental agenda – transportation and energy use. Transport and energy policy are obviously connected and need to be planned in an integrated fashion. Transport environmental policy must be considered in an ever widening context. Communications, agricultural, land-use planning and energy policy are all vitally linked with transport policy.

Transport is responsible for a wide array of environmental problems and the list is growing. Problems associated with all phases of the transport life cycle (manufacture, use and post-use) are increasingly scrutinized and are becoming the subject of public policy. The definition of problems associated with transport is expanding beyond conventional air pollutants such as NO_x and particulate matter to include quality of life issues such as noise and congestion. The trends in vehicle miles travelled and the resulting congestion are being quantified in terms of decreased economic productivity.

Policies to address this growing list of transport-related environmental problems are also proliferating. Everything from specifying the chemical composition of fuels that vehicles use, to limiting transport use, to specifying what type of transport must be supplied to the market are being implemented through public policy.

Recently, advances in market-based mechanisms such as fuel taxes, road pricing and parking charges, and various financial incentives have significantly expanded the array of policy tools available to governments. The UK Government is currently emphasizing the potential advantages of market mechanisms in its overall policy approach to transport environmental issues. Pricing mechanisms, especially in the area of CO_2 control, are being fully explored. This report does not evaluate the relative worthiness of various policy approaches. It is likely that governments will have to rely on a *variety* of policy approaches – regulation, fiscal incentive and voluntary measures – in order to successfully address the future. For industry, regardless of the policy approach chosen by government, there is ample need for technological innovation to tackle the challenges of the environment.

This second phase of the UK Environmental Foresight Project can by no means address all of the environmental issues associated with transport. Nor can it address all of the solutions and policies being implemented. The report covers both traditional regulatory measures and their future development (eg vehicle emission standards) as well as new innovative policy mechanisms, such as accelerated vehicle scrappage. Despite its wide coverage, this study must be considered one contribution to a complex agenda.

The overall goal of this phase of the Foresight Project is to help define the future UK transport-related environmental agenda, some of the driving factors creating that agenda, and some of the potential policy responses. Figure 1 below indicates the scope of this study and puts it in context with other factors that will shape the UK's transport and environment agenda.

Figure 1: Project Scope

Road Transport & Environment

```
                    ┌──────────────┐
                    │  Transport   │
                    │ Environmental│
                    │   Problems   │
                    └──────┬───────┘
                           │
                           ▼
┌──────────────┐    ┌──────────────┐    ┌──────────────────────┐
│  Transport   │───▶│ Future UK    │◀───│ Related Policy Areas │
│ Environmental│    │  Transport   │    │    - transport       │
│    Policy    │    │ Environmental│    │    - energy          │
│              │    │Policy Agenda │    │    - planning        │
└──────────────┘    └──────▲───────┘    │    - communication   │
                           │            └──────────────────────┘
                           │
                    ┌──────┴───────┐
                    │  Transport   │
                    │   Market     │
                    │   Changes    │
                    └──────────────┘
```

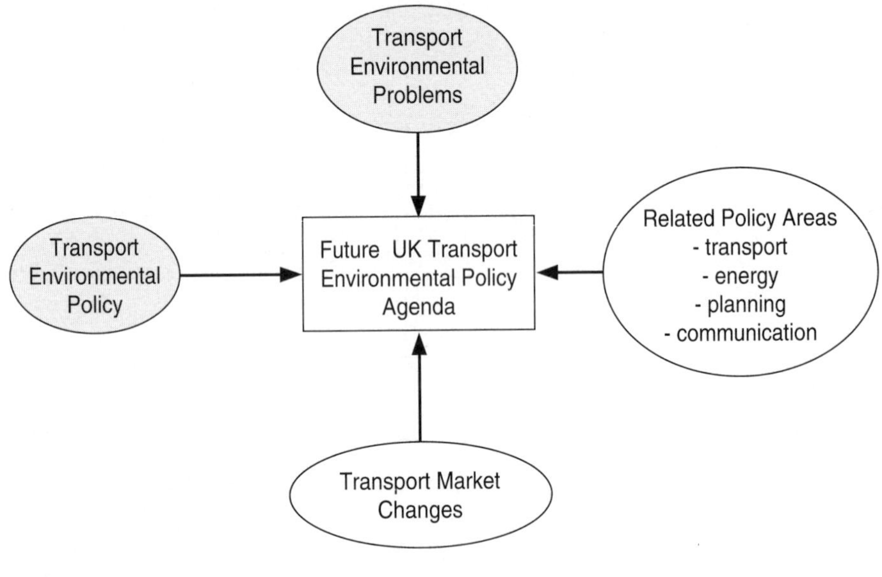

As illustrated, this current study will look at two of the many factors that will be influential in shaping the UK's future transport-related environmental agenda. The first factor will be a look at a limited set of environmental problems emanating from the road transport sector. Air, rail and shipping transport are not considered in this study. The environmental problems addressed in this study are limited to air emissions including carbon dioxide. Noise pollution is addressed in Volume 3 of the report. Other environmental problems, such as hazardous and solid waste or water pollution are not addressed. The study is primarily concerned with the environmental problems associated with the *use* of road transport vehicles, rather than with their manufacture or post-use phases.

The land-use consequences of road transport are not addressed in this report although this is a crucial issue for UK transport policy. Likewise, the use of land planning and other planning methods to alleviate transport problems are not addressed in this report[1].

This volume reviews the contribution of road transport to these environmental problems and how that contribution has changed over time. The review covers both the UK as well as elsewhere. The significance of these pollution problems, particularly on human health, are reviewed in detail. New developments in public health research are described.

The second factor shaping the future agenda concerns the *policies* that have developed to address these various transport environmental problems. The report outlines the changes in road transport policy over time and illustrates future directions for the UK in light of these trends. The report concentrates primarily on policies affecting the individual vehicle from a technical perspective, rather than policies affecting the demand and management of overall transport.

Transport planning is necessary for successfully addressing vehicle environmental problems. This study should be considered a companion to the significant on-going work in transport planning and management. Future transport policy will have to contain both types of measures.

Another factor that will affect a future transport environmental agenda will certainly be technological changes in the vehicle industry. Technological change can either be responsible for the creation of new environmental problems or it can be developed as the solution to existing ones. Transport technological change both drives and is driven by the future environmental agenda. Technological change for the purpose of environmental quality improvement, however, rarely occurs in the absence of strong policy or market drivers. To date, with the exception of fuel efficiency, the market demand for an improved vehicle environmental performance has not been strong enough to take the place of public policy and regulation.

With the exception of vehicle technologies associated with noise reduction reviewed in detail in Volume 3, the vast subject area of vehicle technology development in response to environmental problems is not covered in detail in this report. There are a variety of existing and on-going studies outlining the

---

1 For further information on this subject see: Department of Environment & Department of Transport, 1993, *Reducing Transport Emissions Through Planning*, HMSO, London.

beneficial aspects of new designs of engines, vehicles and vehicle fuels. While these are important in shaping the future agenda, they are for the most part not the driving factor.

This report is directly related to technological development, however, in that it illustrates the future goals and targets that technology will have to address. For example, as health research uncovers previously unknown adverse health effects associated with a by-product of a particular vehicle fuel, and as that adverse effect becomes subject to policy, technological development becomes one of the potential responses. This report covers in detail, then, the first two factors – the future development of environmental problems associated with transport and certain policy mechanisms that may be adopted in response. Concerning the affect of current policy on the future agenda, the report provides a detailed analysis of policy developments outside the UK.

Another limitation of the study concerns costs. For the majority of the policy trends noted, there is little discussion of either the cost of the proposals, an evaluation of their cost-effectiveness, or an evaluation of the cost-effectiveness of other policy options. This omission should not be interpreted as a statement that such analysis is unnecessary. Indeed, it is crucial for governments faced with growing demand for pollution control to implement the most cost-effective policy. This need for cost-effectiveness and cost-benefit evaluation will only increase in the future, as was noted in the national foresight workshop reviewed in Volume 1 of the report.

2 The Motor Vehicle Contribution to Air Pollution in the UK

Road transport contributes to many air pollution problems that are the focus of public concern and government policy. Sulphur and nitrogen oxides contribute to acid rain. Nitrogen oxides and volatile organic chemicals are the precursors of tropospheric ozone or 'smog'. Particulate matter and carbon monoxide adversely affect human health and carbon dioxide emissions are now a concern because of their global warming potential. This chapter assesses the current and future contribution of road transport in the UK to these various air pollution problems. Table 1 highlights road transport compared to other sources of some of these principal air pollutants.

Table 1 Sources of the Principal Air Pollutants In the UK (1990)

Source	% of Total Emissions					
	Sulphur Dioxide	Black Smoke	Nitrogen Oxides	Carbon Monoxide	Carbon Dioxide	Volatile Organic Compounds[2]
Road Transport	2	46	51	90	19	41
Electricity Supply Industry	72	6	28	1	34	–
Other Industry	19	14	9	4	26	52
Domestic	3	33	2	4	14	2
Other	7	1	9	–	7	4
Total (kT)	3,774	453	2,719	6,659	160[1]	2,396

1 Measured in Million tonnes carbon; source "The UK Environment," DOE, 1992.
2 The term Volatile Organic Compounds does not include methane. The evaporation of petrol during production, storage and distribution is included under "other industry". Its evaporation from the petrol tank and carburettors of petrol-engined vehicles is included under "road transport".

Source: Urban Air Quality in the United Kingdom: First Report of the Urban Air Review Group, January 1993

Table 2 compares EC road transport emissions relative to other sources:

Table 2 Contribution of Sources to Air Pollution in the EC (%) 1987[2]

Sector	CO_2	SO_2	NO_x
Energy	37.5	71.3	28.1
Industry	18.6	15.4	7.9
Transport	22.9	4.0	57.7
Others	21.0	9.3	6.3

The Growing Significance of Road Transport Air Pollution Emissions

The contribution of road transport to UK emissions of various pollutants has changed over time (see Table 3 below). The relative contribution varies according to the emission control policy and technology in place at the time, the characteristics of fuels used, the growth and composition of the UK fleet, and changes in the contribution of other sources of air pollution (eg, stationary and industrial sources).

Table 3 Road Transport Contribution to Air Pollution in the UK[3]

(%)	1970	1980	1982	1984	1986	1988	1990	1992	1995	2000	2005	2010
NO_x	26	35	38	42	43	48	51	50[(1)]	49	46	40	37
CO	61	80	83	87	82	86	87					
NMHC	32	35	34	32	34	38	41	36[(2)]	33	33		
Black Smoke	10	21	22	27	26	35	44					
Lead				88	74	76	69					
CO_2	10	15	16	17	16	18	19	19[(3)]	19	20	20[(4)]	

1 1992 through 2010 emission estimates are approximate midpoint percentages from projected high and low emission scenarios (see Eggleston, 1992).
2 1992, '95 and '00 figures are from UK VOC Control Discussion Document.
3 High CO_2 emission projection scenario is used for '92, '95, '00. (see Eggleston, 1992).
4 From UK DoE Climate Change Discussion Document, excluding buses.

2 European Commission, 1990, "DG XVII: Energy for a New Century – The European Perspective", in Research and Technology Strategy to help overcome the environmental problems in relation to transport: Commission of the European Communities Science Research and Development (Monitor), Overall Strategy Review; Brussels, 1992.

3 Sources: Department of the Environment, 1992, *The UK Environment*, Government Statistical Service, HMSO, London.
Quality of Urban Air Review Group, 1993, *Urban Air Quality in the United Kingdom*: First Report.
H S Eggleston, 1992, *Pollution in the Atmosphere: Future Emissions from the UK*, Warren Spring Laboratory, LR888 (AP), Stevenage.

Road Traffic Growth

Obviously the growth in road traffic (vehicle miles driven) is the major contributor to all of the trends discussed above. From 1949–1989, average growth in vehicle miles travelled in the UK was 5.5% annually and **770%** in total. Car traffic increased **16 fold** to over 7% per year and HGV growth in vehicle miles travelled was less than 2.5% per year.

2.1 UK Road Transport Air Pollution Trends in Comparison

To put the UK's road transport air pollution contribution in perspective, Table 4 gives similar information on the contribution of road transport to air pollution in the US over the same period.

Table 4 Road Transport Contribution to Air Pollution in the US (%)

	1970	1980	1982	1984	1986	1988	1990	1992	1995	2000	2005	2010
NO_x	34	38	38	35	33	31	29	28	24	19	18	16
CO	64	70	63	61	60	52	50	48	43	40	44	44
VOC	36	33	36	32	32	29	27	23	19	12	12	12
PM	5	13	15	15	16	16	17					
Lead	76	79	80	83	39	32	28					

Sources: "National Air Pollutant Emission Estimates: 1940–1990", US Environmental Protection Agency, 1991 and "National Air Quality and Emission Trends Report, 1990", US EPA, 1991

Table 5 and Figures 2 to 5 compare the trends in road traffic air pollution emissions in the UK and US from 1970 to 1990. The reader should note that there are two factors responsible for the individual country trends and differences. The first is the rate of vehicle traffic growth over time – specifically, the rate of growth of vehicle types and their corresponding fuel usage. The second more important factor is the implementation of various air pollution control measures (either emission control vehicle technologies or fuel quality specifications).

Table 5 Percentage Changes in UK[4] and US[5] Road Transport Vehicle Emissions (by weight)

Pollutant	% Change from 1970–1980 UK	US	% Change from 1980–1990 UK	US	% Change from 1970–1990 UK	US
NO_x	+32	+25	+67	−41	+126	−12
CO	+40	−34	+46	−60	+104	−115
VOC	+31	−18	+12	−51	+47	−78
PM	+16	+22	+82	+18	+113	+44
SO_2/SO_x	−6	+33	+51	+50	+42	+100
CO_2	+30	na	+40	na	+81	na

The trends in UK and US air pollution of major mobile source conventional pollutants are quite clear.

Figure 2: Road Transport Contribution to UK & US NO_x Emissions 1970-1990

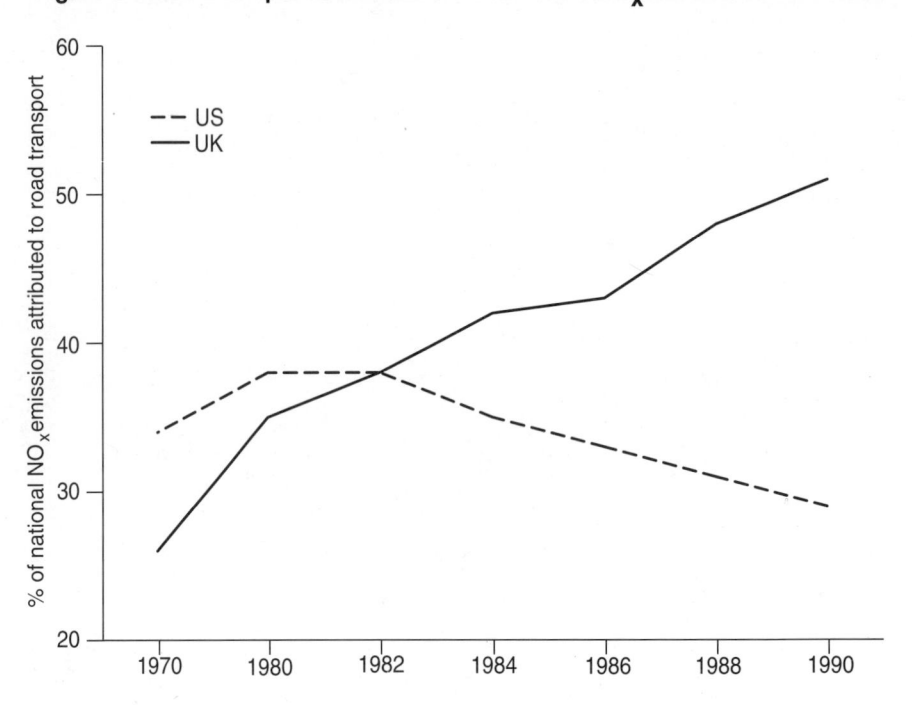

---

4 H S Eggleston, 1992, *Pollution in the Atmosphere: Future Emissions from the UK*, Warren Spring Laboratory, LR 888(A), Stevenage.

5 US Environmental Protection Agency, 1991, *US Emissions: National Air Pollutant Emission Estimate 1940–1990*, Office of Air Quality Planning and Standards, Research Triangle Park, NC.

**Figure 3: Road Transport Contribution to
UK & US Carbon Monoxide Emissions 1970-1990**

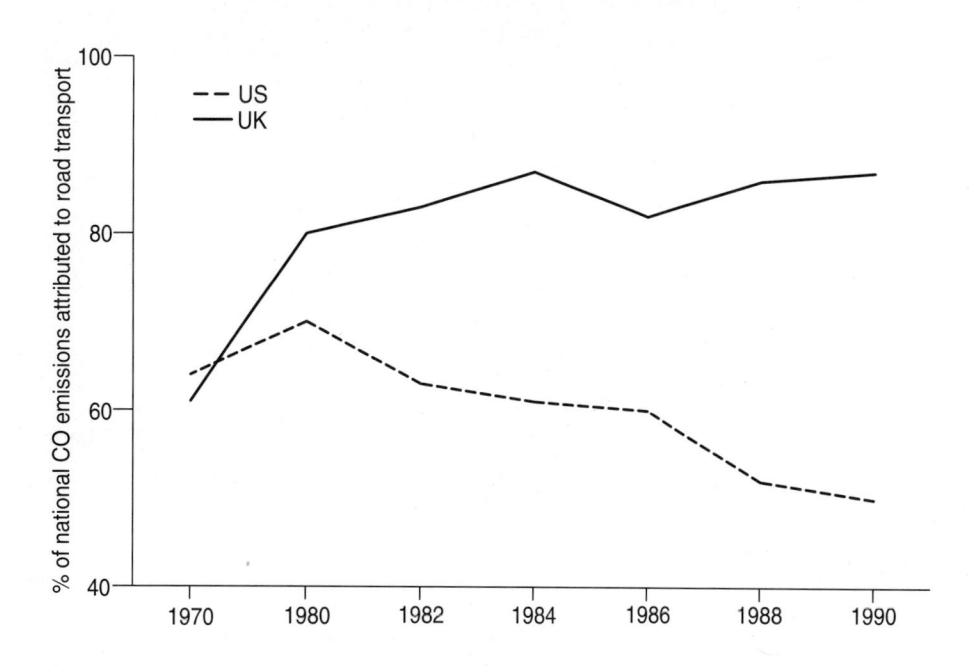

**Figure 4: Road Transport Contribution to
UK and US HC Emissions 1970-1990**

**Figure 5: Road Transport Contribution to
UK & US Particulate Matter Emissions 1970-1990**

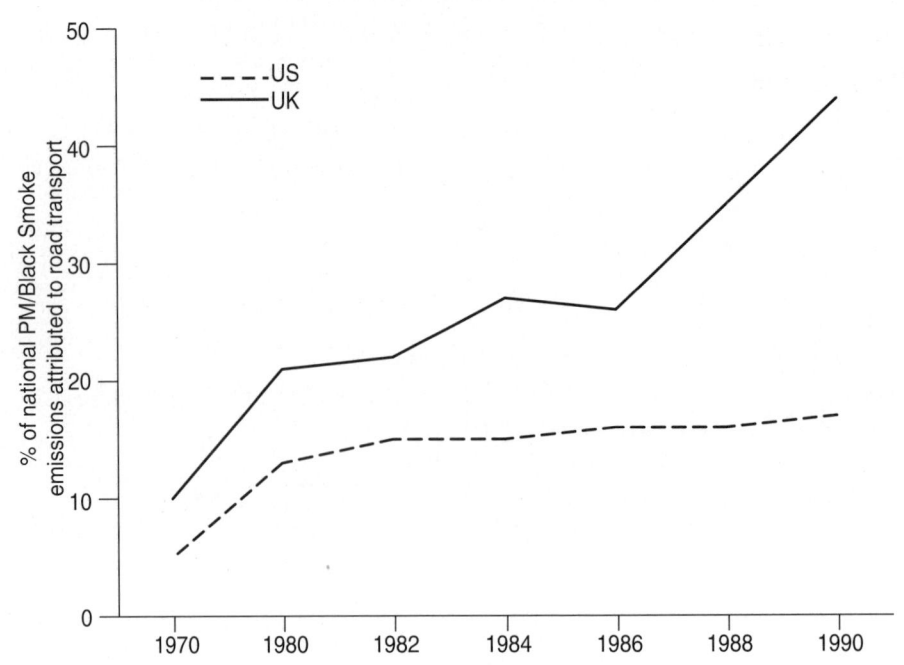

The differences in relative contribution for a given year and trends in relative contribution over time are due to a multitude of factors. The earlier and greater decrease in the relative contribution of motor vehicle lead emissions in the US, for example, is due to earlier restriction of lead in petrol in the US and broader market penetration aided by bans on leaded petrol and widespread introduction of the catalytic converter. The relative greater contribution of transport to UK PM (black smoke) levels is due to a greater percentage of the road transport fleet operating on diesel and a relative lack of diesel emission exhaust standards.

This comparison should not be used to interpret the advantages or disadvantages of transport-directed air pollution control policies, since the air pollution problems in the US and the UK are certainly not identical. It does, however, indicate the early progress made in the US compared to the UK. It also illustrates the potential that additional policies in the UK have in terms of reducing the relative contribution of road transport to overall levels of air pollution.

The real utility of such a comparison lies in examining the policies and regulations behind the emission reduction trends in the US and assessing their relevance for the UK.

2.2 The Significance of Road Transport Emissions in Urban Environments

Ambient air pollution reduction policies can be developed with several scopes in mind. The focus can be on national programmes, rural programmes, urban programmes and a global level. In addition, particular 'air sheds', (eg, tropospheric ozone transport regions) can be isolated as the focus for control programmes. In order to implement programmes of different scope, information must be assembled with that goal in mind.

Figures 6 to 8 indicate the relative difference in ambient concentrations of pollutants between London and the UK as a whole. Such information provides indications of key source inventory differences that must be addressed if policy is to be focussed on urban pollution as opposed to overall national ambient levels. For example, industrial contributions to NO_x levels are dominant if an overall UK perspective is taken, but if control of NO_x levels in urban areas is a policy goal, then the transport sector bears closer scrutiny. Likewise, domestic sources of black smoke are dominant at a national level in the UK, but the transport sector dominates within urban settings such as London.

2.3 Vehicle Type and Road Transport Emissions

In order to optimise pollution control policies, the relative contribution of different types of motor vehicles is important. The following sections describe the contribution of passenger cars and heavy goods freight vehicles to road transport emissions.

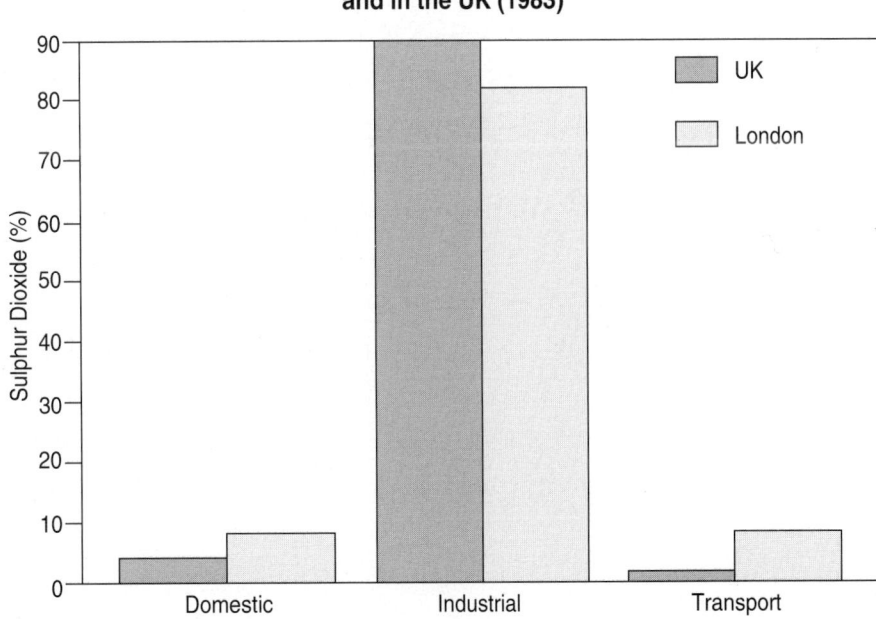

Figure 6 : Sources of SO_2 in Greater London (April 1983-March 1984) and in the UK (1983)

Source: Urban Air Quality in the UK, First Report of the Urban Air Review Group, London, 1993

Figure 7 : Sources of NO$_x$ in Greater London (April 1983-March 1984) and in the UK (1983)

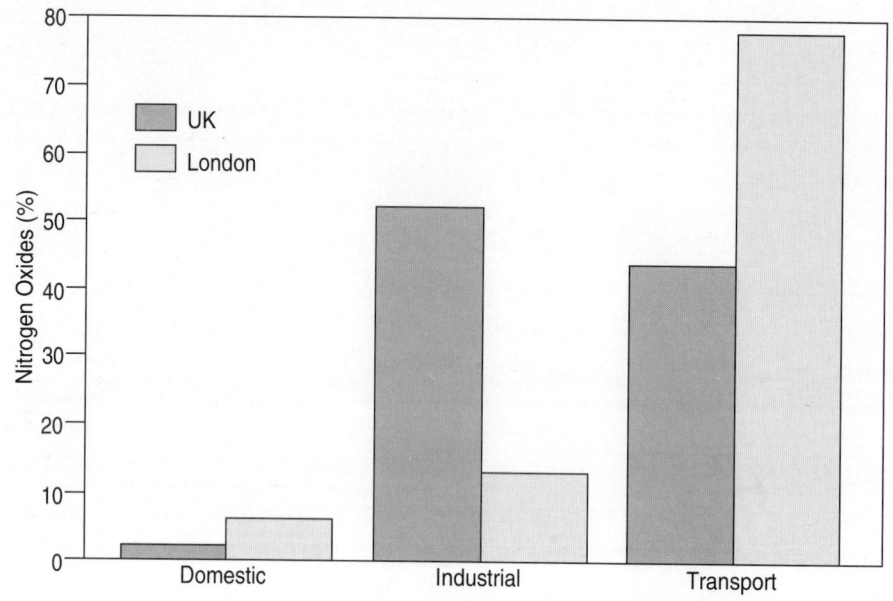

Source: Urban Air Quality in the UK,
First Report of the Urban Air Review Group, London, 1993

Figure 8 : Sources of Black Smoke in Greater London (April 1983-March 1984) and in the UK (1983)

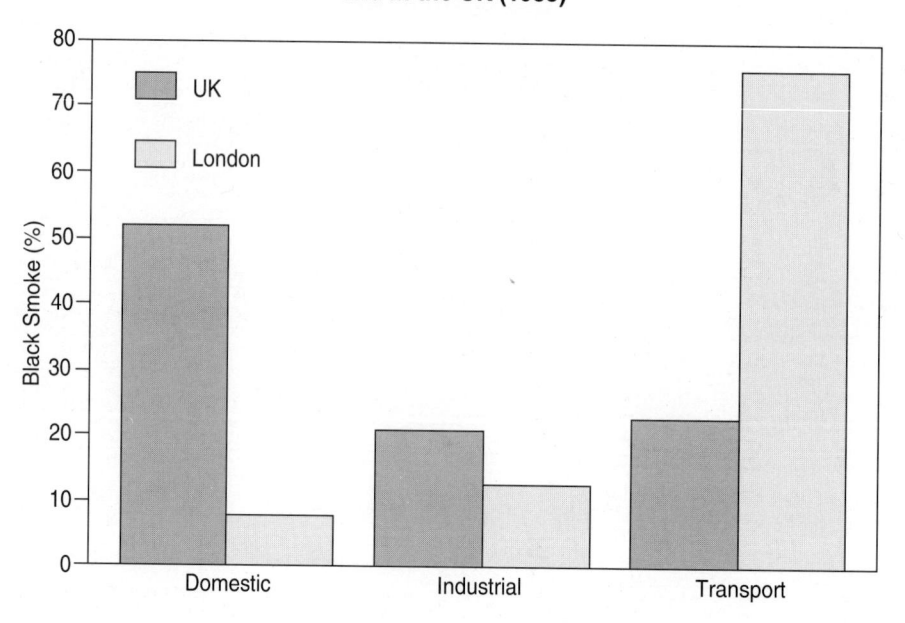

Source: Urban Air Quality in the UK,
First Report of the Urban Air Review Group, London, 1993

2.3.1 Contribution of Passenger Cars to UK Road Transport Emissions

Passenger cars (as opposed to HGV and buses) are major contributors to the total emissions from road transport (Table 6).

Table 6 Passenger car contribution to total UK road transport emissions (1990)

Pollutant	% from Passenger Cars
Carbon Monoxide	93
Volatile Organic Compounds	61
Nitrogen Oxides	57
Particulates	6
Sulphur Dioxide	39
Carbon Dioxide	69

Source: Pollution in the Atmosphere: Future Emissions from the UK, Eggleston, 1992

2.3.2 Contribution of Heavy Goods Vehicles (HGV) to UK Air Pollution

HGVs (lorries and buses) will take on even greater importance in future transport air pollution control strategies (Table 7). Even with the emission standards scheduled to come into existence in the EC in 1995 for further control of HGV emissions, HGV's percentage contribution to total transport air pollution loads will be increasing. This is especially the case with NO_x.

Table 7 Freight (HGV) % Contribution to Total UK Transport Air Pollution Emissions

	1990	2000	2010
PM	67	66	66
VOC	14	12	17
NO_x	34	37	48

Source: Pollution in the Atmosphere: Future Emissions from the UK, Eggleston, 1992.
Note: VOC emission estimates excluding evaporative emissions

The growing importance of HGV air pollution emissions in comparison to overall road transport emissions is illustrated in Figure 9 below. As NO_x emissions from passenger cars become relatively well controlled with the advent of the catalytic converter and other emission control policies, the NO_x emissions from the freight sector take on a greater relative importance – even though absolute levels of road transport NO_x emissions are much lower in the future than today's levels. Thus, if NO_x is still an air pollution problem after the turn of the century, HGV will increasingly be the focus of environmental policy.

Figure 9: NO$_x$ Emissions from Road Transport in the UK

Low Traffic Growth Assumption

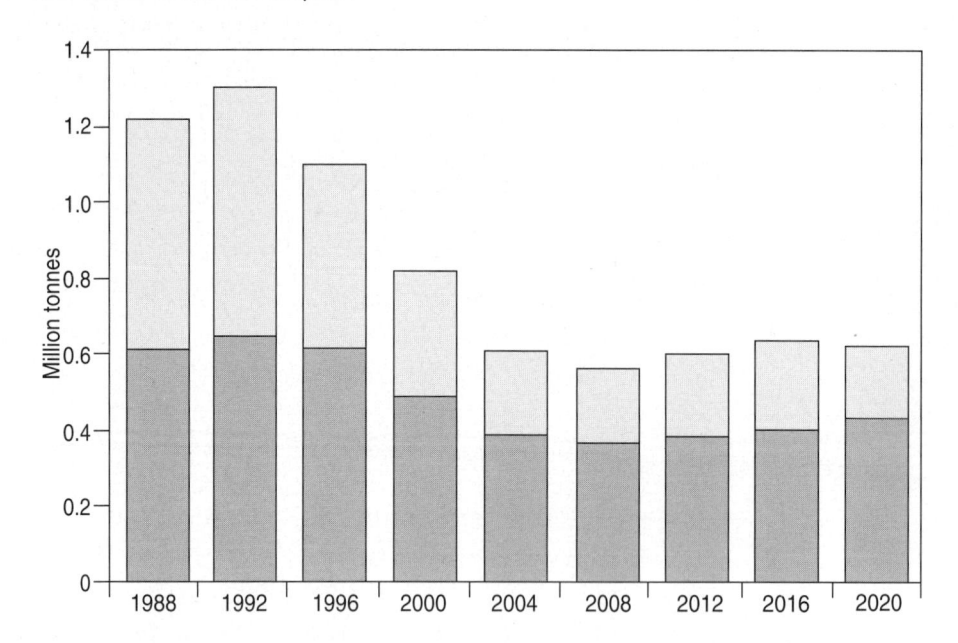

High Traffic Growth Assumption

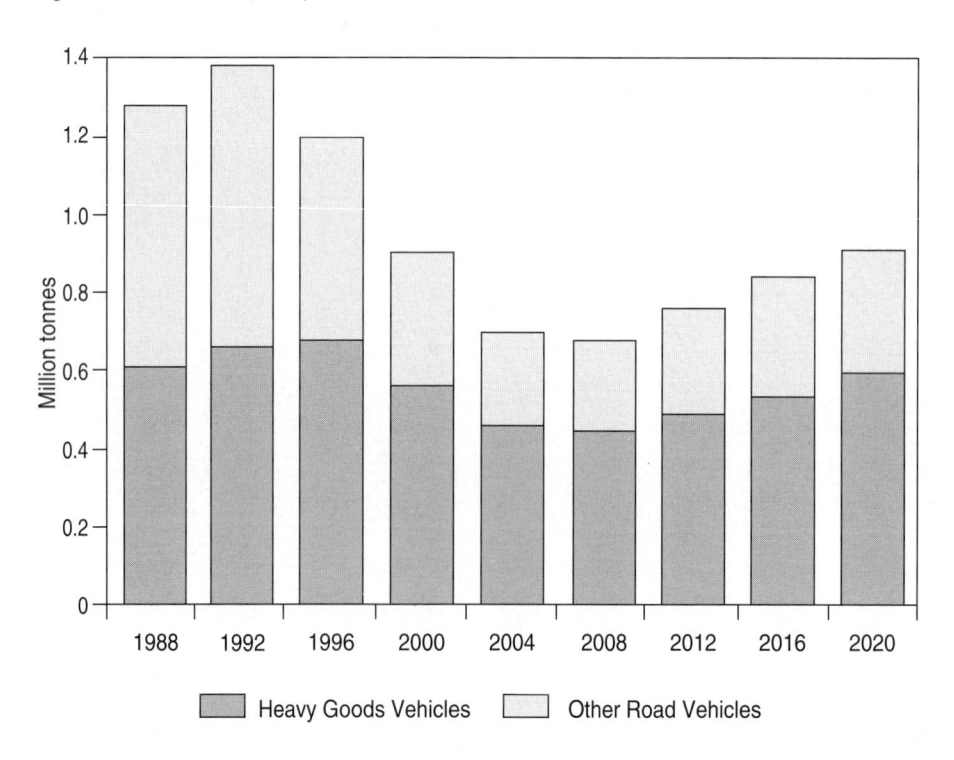

■ Heavy Goods Vehicles   □ Other Road Vehicles

Projections of nitrogen emissions on the basis of two forecasts of demand growth issued by the Department of Transport 1989: low forecast on top and high below. The projection takes into account:
- implementation of the agreed amendments to EC Directive 88/77 EEC in 1993 and 1996;
- replacement and growth of the fleet of HGV and passenger service vehicles (over 32 seats) as forecast by the Department of Transport.

Source: Emissions from Heavy Duty Diesel Vehicles,
Royal Commission on Environmental Pollution 15th Report,
London HMSO, September 1991

HGV emissions also play an important role in the US. In California for example, the California Air Resources Board estimates that by the year 2000, heavy duty vehicles will contribute more than 50% of the road transport NO_x emissions and more than 84% of the particulate matter[6].

2.4 The Contribution of Fuels to UK Road Transport Air Pollution Emissions

The fuels that are used by road transport and other sectors are a major source of ambient air pollution. The Quality of Urban Air Review Group's recent report "Urban Air Quality in the United Kingdom" details the following contributions of diesel and petrol transport fuels to total UK emissions:

Table 8 % of Total Contributions to UK Emissions

	SO_2	BS	NO_x	CO	VOC
Petrol	1	3	29	87	27
Diesel	1	42	21	3	7

Source: DOE, 1992, Digest of Environmental Protection and Water Statistics No.14, 1991

The relative contribution of petrol and diesel-fuelled vehicles to the total UK road transport emissions over time for NO_x and CO_2 is depicted in Figures 10 and 11 below.

2.5 Road Transport Contribution to UK VOC Emission Inventories

Volatile organic chemicals are composed of a number of specific chemicals and classes of chemicals. They are emitted from a variety of sources such as motor vehicles, solvent usage, oil refining, petrol storage and other industrial processes.

As part of its compliance strategy with the UN ECE VOC long-range transport protocol, the UK will require a 30% reduction in the total VOC emissions inventory from a base year of 1988 by 1999. These reductions are being sought to reduce tropospheric ozone or 'smog'. Because many of the chemicals classified as VOCs have other ecological and human health effects or may photochemically transform themselves to more hazardous air pollutants, the reduction plan will have the added benefit of reducing contributions to these problems, too.

---

6 M P Walsh, 1993, *Car Lines, International Regulatory Developments: The Year in review*, volume 10 number 4, Arlington, VA.

Figure 10: UK Road Transport Emissions of NO$_X$ (1970-1990)

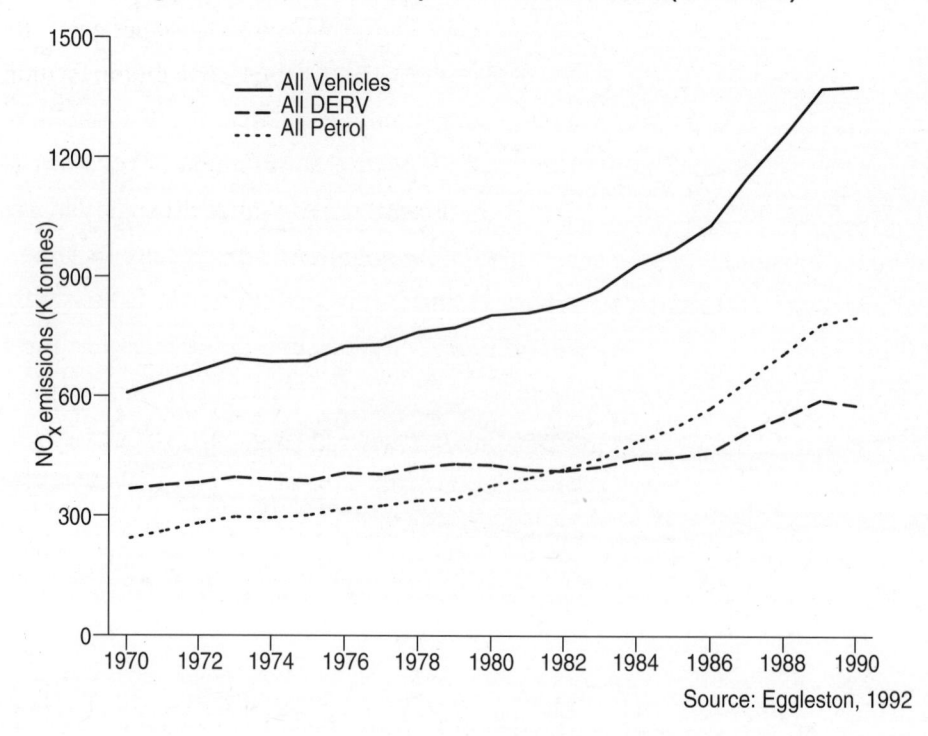

Source: Eggleston, 1992

Figure 11: UK Road Transport Emissions of CO$_2$ (1970-1990)

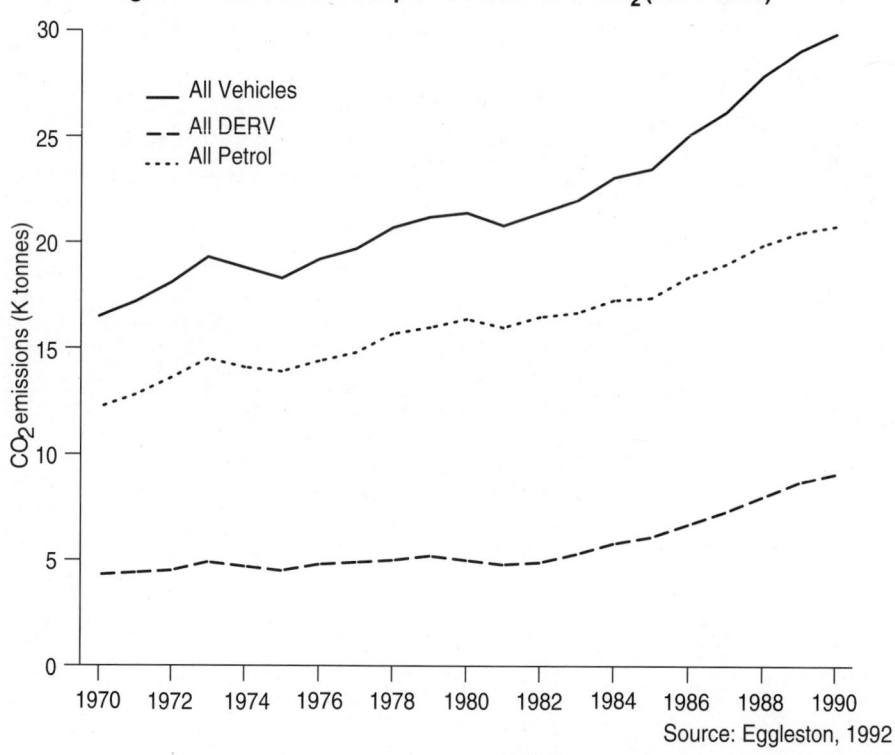

Source: Eggleston, 1992

- In 1988, total UK VOC emissions were 2,734 (kt) and mobile sources contributed 972 or 36% to that total[7]. Petrol exhaust, at 644 ktonnes, is the largest single contributor (stationary or mobile) to the VOC emission inventory.

- The relative contribution of transport to the total VOC inventory is not projected to change dramatically in the future. Projected trends in VOC emission inventories are included in Chapter 3 below.

In addition to the relative importance of road transport to VOC emission inventories based on volume, it is important to look at the type of VOCs emitted by transport in terms of their capacity to form ozone. Not all VOCs have the same ozone formation potential. Those emitted by road transport have a relatively high photochemical oxidant production capability. For purposes of ozone control then, transport emission reductions should be given higher priority than less reactive VOCs and sources.

2.6 Road Transport Contribution to UK CO_2 Emissions

UK CO_2 emissions peaked at 181 million tonnes in 1979 and currently stand at 160 million tonnes in 1990. Table 9 below indicates the trends, according to source, over the time period 1970 to 1990. Future projections are also included.

Table 9 Past trends and projections of UK CO_2 emissions by sector (MtC)

	1970	1980	1990	2000	2010	2020
Transport	23	29	38	45	49	62
Industry	85	64	56	58	61	71
Commerce	10	11	14	18	30 (with public sec)	45 (with public sec)
Public Sector	10	12	10	9		
Households	54	48	41	41	42	42
TOTAL	182	165	160	170	183	221

Source: Climate Change: Our National Programme for CO_2 Emissions
A Discussion Document, UK DoE, December 1992, London

Road transport, as the end-user of energy and emitter of CO_2, contributed 21% to total UK CO_2 in 1990 as Figure 12 indicates. In 1970, road transport contributed approximately 9% of total UK CO_2 emissions and in 2005 it is estimated to contribute over 23% to the UK total.

[7] Department of the Environment, 1992, *Reducing Emissions of Volatile Organic Compounds (VOCS) and levels of ground level ozone: A UK Strategy*, Draft Consultation Document, London.

Figure 12: Carbon dioxide emissions, by end user[1], 1990, UK

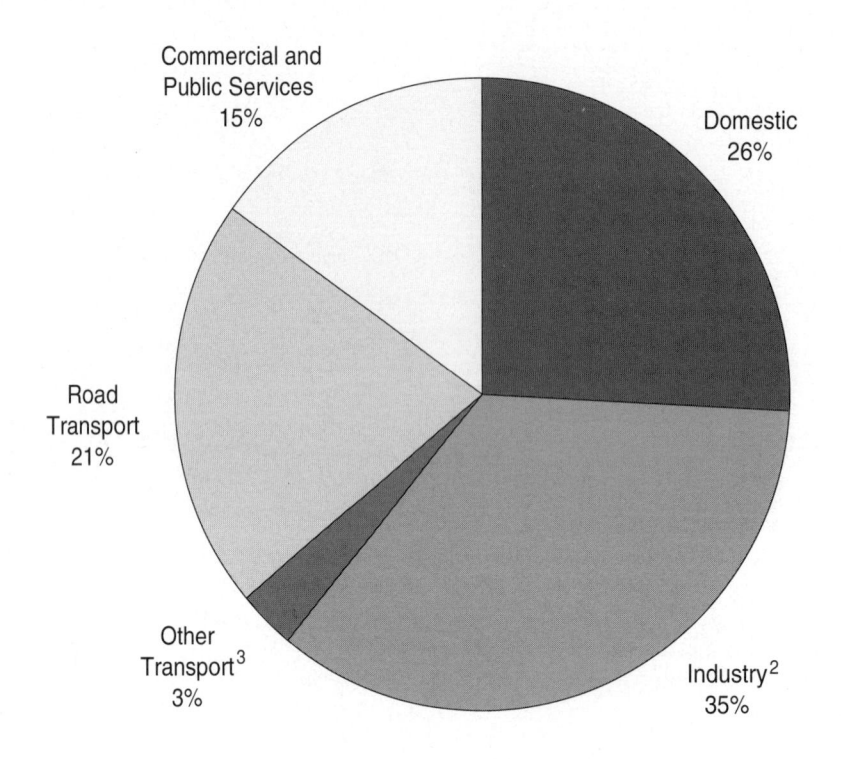

- Domestic 26%
- Industry[2] 35%
- Other Transport[3] 3%
- Road Transport 21%
- Commercial and Public Services 15%

Notes:
1 Figures for end users include emissions from power stations which have been allocated to the various sectors that use the electricity generated. They also include allocations of emissions from other primary and secondary fuel producers.
2 Includes agriculture (1%).
3 Coastal shipping (<12 miles) railways, civil aircraft (on ground, landing and take off up to 1km).

Source: The UK Environment, DoE, London, HMSO

3 UK Ambient Air Quality Concentrations and Trends to Date

A short review of available evidence and studies of air pollutant ambient trends in the UK will help provide a context within which to examine the necessity for and impact of proposed air pollution control measures and policies. Past air quality trend information varies significantly according to the pollutant examined. Projected future trends in road transport emissions are included at the end of this chapter.

3.1 Nitrogen Oxides Concentrations and Trends to Date

The recent UK Photochemical Oxidants Review Group[8] summarized atmospheric concentrations and trends of nitric oxide and nitrogen dioxide in the following manner:

- Urban concentrations of both chemicals are greater than rural; city centres exhibit the highest concentrations; and winter mean concentrations of NO_x and NO_2 are significantly higher than summer mean concentrations.
- Annual mean concentrations have nearly doubled at some rural sites from 1979–1987.
- No clear evidence of upward trends at urban sites is noticeable, with the exception of the results from one rooftop Central London monitoring site.
- Urban background sites exhibit hourly mean NO_2 concentrations in the range of 100–400 ppb.
- Roadside concentrations of both pollutants are elevated over urban background levels and are directly attributable to the presence of motor vehicle traffic.
- The highest level NO_2 episode measured in the UK was in London in December 1991. High NO_2 episodes were repeated the following year in Manchester, where a 369 ppb peak was recorded.

---

8 UK Photochemical Oxidants Review Group, 1990, *Oxides of Nitrogen in the United Kingdom, Second Report*, London.

3.2 Tropospheric Ground-Level Ozone Concentrations and Trends to Date

Tropospheric ozone (O_3) or smog is a secondary atmospheric pollutant formed from the primary emissions of volatile organic chemicals or hydrocarbons and nitrogen oxides. Although trends in nitrogen oxide concentrations over time are available (as summarized above), trends in average VOC or hydrocarbon trends are not readily available. Trends in ozone are available, however. Accumulated evidence indicates that:

- Annual mean and 98th percentile hourly mean ozone concentrations at urban sites are lower than at nearby rural and suburban sites[9].
- Peak ozone episodic statistics and extreme situations of ozone concentrations do not duplicate this rural-urban distribution. Often peak episode measurements are in urban areas. This may be due to the relatively small size of the ozone measurement data base in the UK.
- Urban sites show a reduction in O_3 concentrations by mid-morning, which is most likely a result of NO scavenging caused by elevated emissions of NO during morning rush hours.
- Elevated rural ozone levels are attributed to the chemical transformation of urban-generated HC and NO_x, as plumes disperse from urban locations and become transformed.
- Statistical analysis of the monthly mean and maximum ozone concentrations in London from the period of 1972–1990 indicates that the weighted monthly mean at London Victoria decreased in recent years under the influence of increased NO_x emissions. This is expected to reverse itself as NO_x becomes relatively controlled.

3.3 Particulate Matter Concentrations and Trends to Date

- Black smoke[10] concentrations in urban UK locations have steadily declined over the past 25 years as shown in Figure 13 below.
- In terms of aggregate amount, emissions have fallen steadily to 453,000 tons since 1986.
- Emissions from domestic sources have declined by 50% between 1980 and 1990.
- For the first time, emissions from diesel fuel and road transport in general were greater than domestic sources in terms of overall emissions, with an almost doubling of diesel emissions from 1980–1990.

---

9 Quality of Urban Air Review Group, 1993, *Urban Air Quality in the UK: First Report*, prepared at the request of the UK Department of the Environment.

10 Smoke being defined here as "suspended particulate of less than 15 µg diameter arising from the incomplete combustion of fuel."

Figure 13: Mean Urban Concentration of Black Smoke in the UK (1962 - 1990)

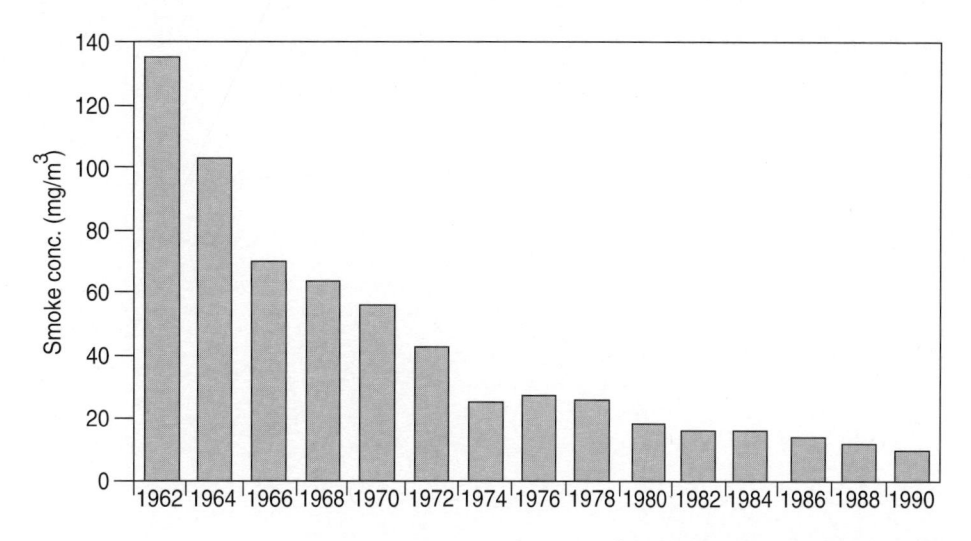

Source: Urban Air Quality in the UK
First Report of the Quality of Urban Air Review Group, London, 1993

3.4 Carbon Monoxide Concentrations and Trends

- Long term trends of ambient carbon monoxide levels in the UK are difficult to discern due to limited monitoring information. One site, Victoria in central London, lends itself to a trends analysis.
- Analysis of reliable data indicates an annual increase of 0.04 ppm from 1980 to 1991[11].
- In 1991, this corresponds to an annual concentration increase of 3% (± 2%) and a similar 3% increase in the 98th percentile concentration amount. Caution should be used in ascertaining ambient CO trends due to severely limited and differential data.
- A diurnal variation indicates a strong correlation between elevated urban CO levels and morning and afternoon rush hour traffic.
- Winter concentrations of CO (most likely due to increased cold-start vehicle emissions) are greater than summer time concentrations.
- Potential increases in average urban CO concentrations could be attributed to the 32% UK CO emission increase over the same 1980–1991 period.

3.5 Lead Ambient Concentrations and Trends

- Lead concentrations, averaged over five sites, have declined by 77% from 1976–1988 (see Figure 14).
- Notable in this trend is the drop in ambient lead levels from 1985 to 1986, after the reduction in the maximum lead content of petrol from 0.4 to 0.15 grams/ml.
- Ambient lead levels decline as distance from sources increases.

---

[11] op cit note 9.

**Figure 14: Annual Mean Concentrations (ng/m³) of Lead in UK Urban Air
(1976 - 1988)**

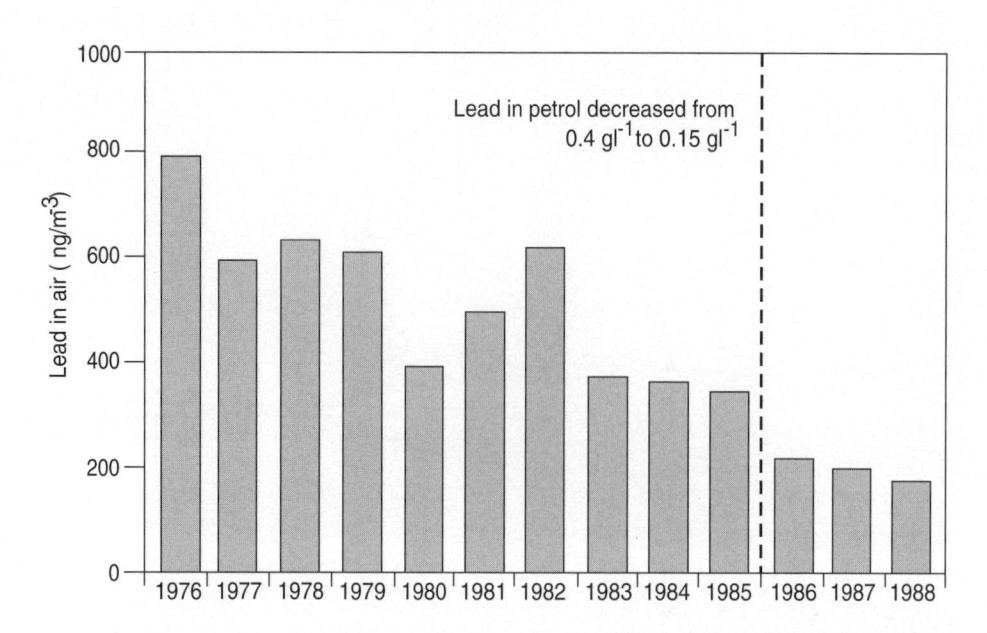

Source: Urban Air Quality in the UK,
First Report of the Quality of Urban Air Review Group, London, 1993

3.6 Future UK Road Transport Emission Projections

By updating an earlier model, researchers at Earth Resources Research have made projections of future UK passenger car air pollution emissions[12]. The work indicates a dramatic reduction in UK passenger car emissions primarily as a result of the introduction of the 3-way catalyst into the UK fleet. From a 1993 baseline, emissions of CO will be reduced by 43%, VOCs by 60% and NO_x by 64% in the year 2003, ten years after industry-wide catalyst use. Figures 15–18 below present emission projections for passenger cars to the year 2025.

---

12 C Holman, J Wade & M Fergusson, 1993, *Future Emissions from Cars 1990 to 2025: The Importance of the Cold Start Emissions Penalty*: A report for World Wide Fund for Nature, UK, Earth Resources Research, London.

Figure 15: UK Passenger Car VOC Emissions Projections 1990-2025

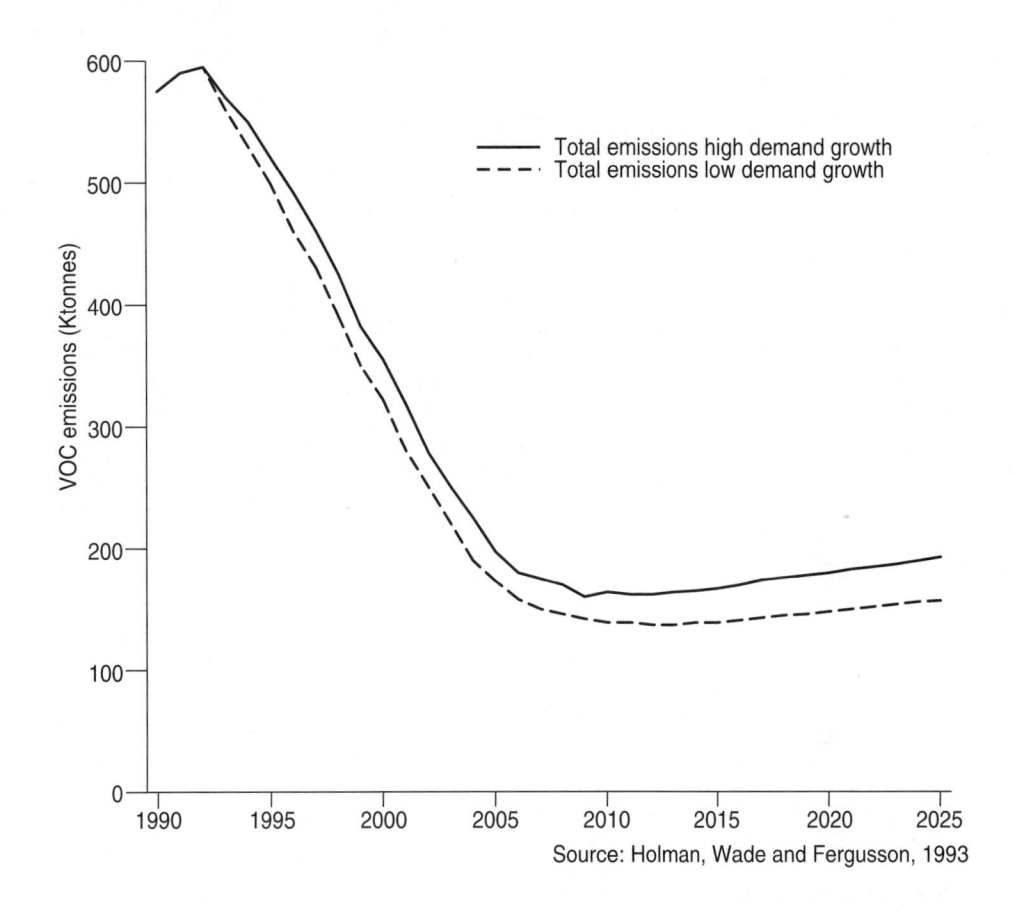

Source: Holman, Wade and Fergusson, 1993

Figure 16: UK Passenger Car NO$_x$ Emissions Projections 1990-2025

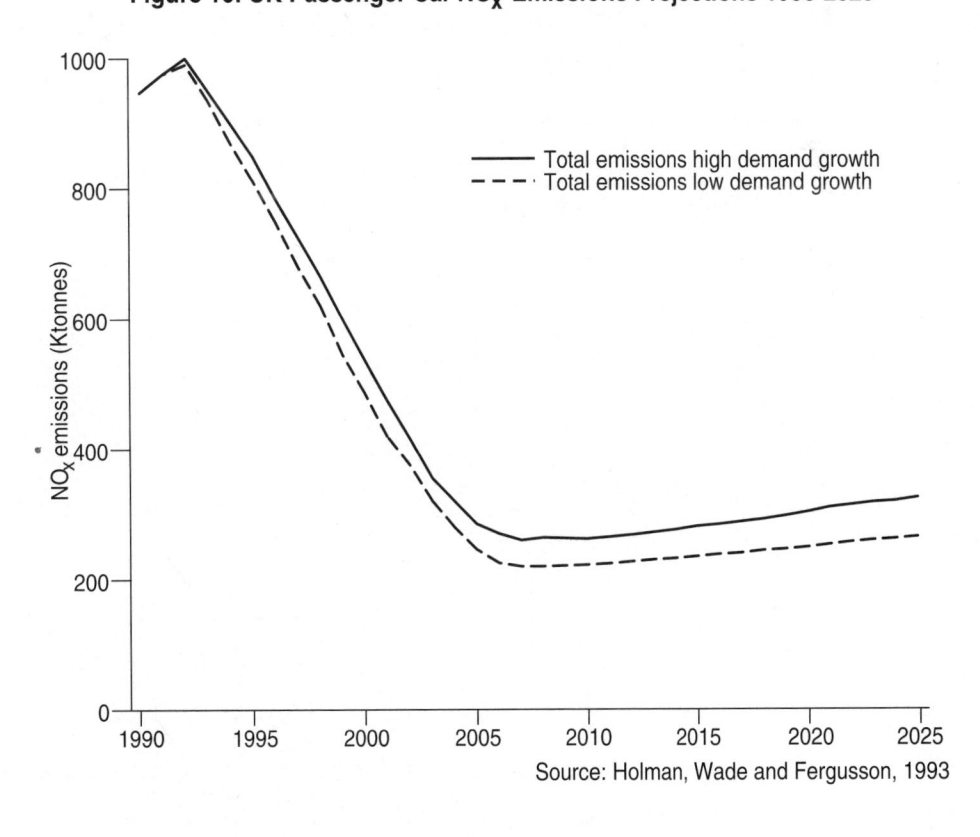

Source: Holman, Wade and Fergusson, 1993

Figure 17: UK Passenger Car CO Emissions Projections 1990-2025

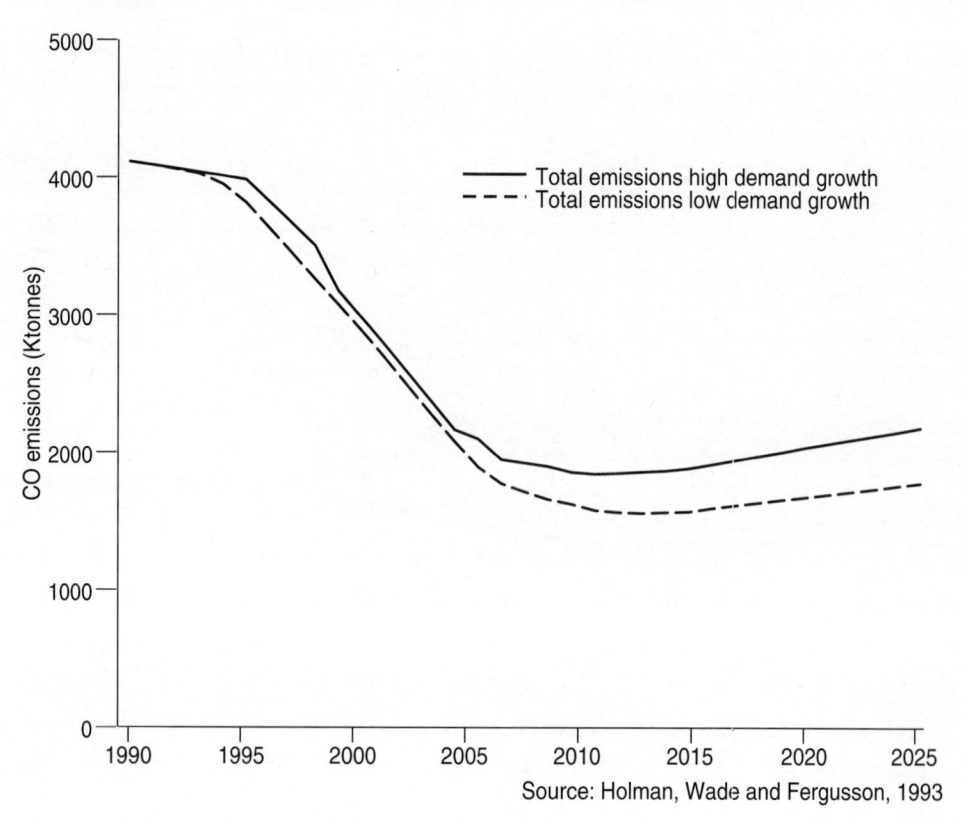

Source: Holman, Wade and Fergusson, 1993

Passenger car carbon dioxide emissions do not favour as well as conventional pollutants. Without substantial progress made in vehicle fuel efficiency (ie miles per gallon), CO_2 emissions are projected to rise as shown in the following figure[13].

---

13 op cit note 4.

Figure 18: UK Passenger Car CO_2 Emissions Projections 1990-2010

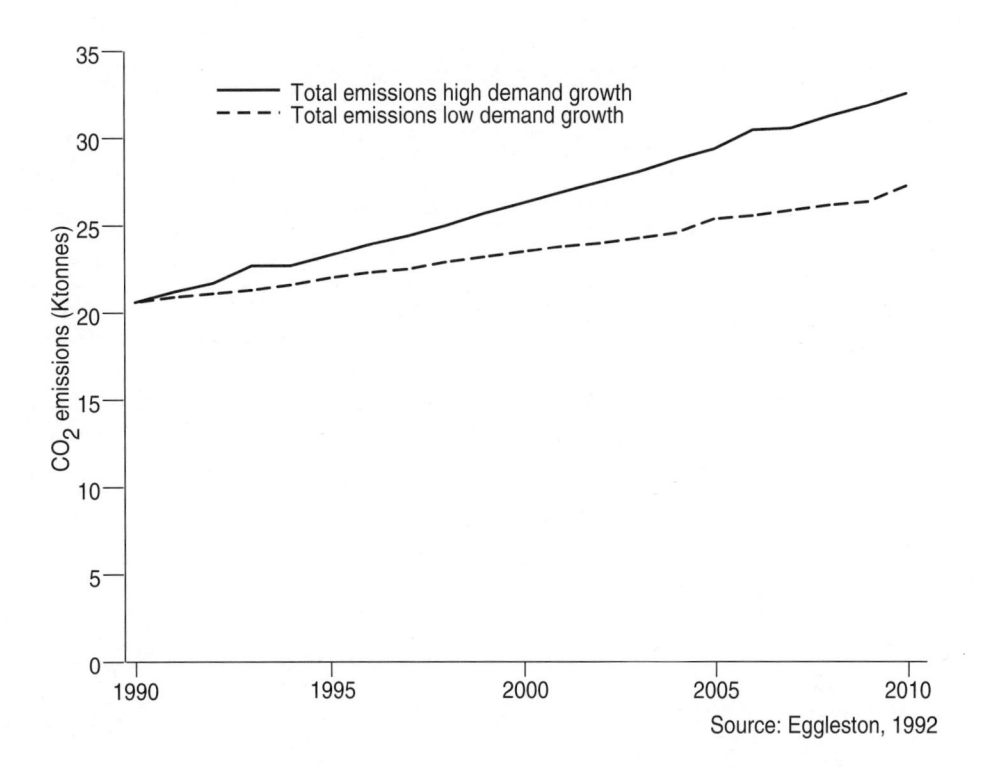

Source: Eggleston, 1992

In addition to the catalyst-related decline, the projections also indicate a slow but steady rise in emissions from approximately the year 2010 onwards. This is due to continued growth in the size of the fleet, vehicle miles driven by the fleet and the difficulty of gaining significant incremental reductions in a catalyst-controlled fleet.

In addition, this research indicates the importance of controlling so-called 'cold start' emissions. Cold start emissions are emitted from vehicles during the start-up phase before the catalyst has had time to warm up and properly eliminate pollution. Approximately 50% of VOC and CO emissions from the UK passenger car fleet will be due to these early emissions by the year 2000[14]. The cold start phase NO_x emissions are relatively less important, amounting to roughly 5% of total passenger car NO_x emissions in the year 1990. The importance of cold start emissions must not be underestimated. Because of the high percentage of very short distance journeys by cars (especially in urban settings), and the high emissions for the first few minutes of car operation, cold start emissions represent a dominant part of the passenger car emission 'portfolio'. Their control has been recognized as a primary goal for UK and EC policy.

14 op cit note 12.

Implications for the UK

- 3-way catalysts will be very successful in reducing individual vehicle and overall fleet emissions in the UK. These quantified source reductions must now be translated into estimates of ambient air quality improvement.

- In the future, with the inevitable rise in fleet emissions and potential degradation in air quality, other types of emission control policies will have to be implemented.

- Because the incremental cost of these future vehicle emission controls will be greater than the incremental cost of reductions gained through catalyst applications, policy makers and manufacturers will have to examine a whole range of new policies. A selection of potential next steps is reviewed later in this volume.

- It is likely that the environmental goals of vehicle emission control policies will change in the future. Toxic air emissions and global pollutants will become a source of concern, requiring new vehicle and fuel policies. In addition, it is likely that the existing ambient standards for conventional pollutants will be made more stringent in the future. These developments, which are described in detail in this volume, will result in a continued pressure to eliminate environmental pollution from road transport.

- Despite the improvements gained through a catalyst-equipped passenger car fleet, it is inadvisable for the UK to relax its efforts to investigate the next generation of vehicle emission control technology and policy. Many air pollution problems will continue to plague the UK.

For example, it is likely that, even with proposed action to limit NO_x and HC from road transport and from major stationary sources, the UK will breach the WHO ozone guidelines in the future. Episodes of 'poor' ozone air quality across southern Britain during summer months can be expected until well into the next century[15].

---

15 Department of Transport & Department of Environment, 1992, *Joint Memorandum to the Royal Commission on Environmental Pollution's Transport and Environment Study.*

4 UK Air Pollutant Compliance with Standards, Guidelines and Directives

Air Quality Criteria

UK ambient concentrations of air pollutants can be compared to a number of air quality criteria:

EC Limit and Guide Values have been set for a number of air pollutants (smoke and sulphur dioxide, nitrogen dioxide, lead, and ozone). Limit values are mandatory and guide values are to provide guidance to EC member states.

World Health Organization air quality guidelines have been developed for 28 substances. They are based on human health criteria, and are intended to provide background information and guidance to governments making human health risk management decisions.

UK Public Information Air Quality Criteria have been established to assist the public in interpreting the public health significance of various ambient concentrations of air pollution. To date, these informational criteria have been determined for SO_2, NO_2 and O_3 and correlate ambient concentrations to ratings of 'very good', 'good', 'poor', and 'very poor'.

Summaries of compliance with the various EC directives and WHO Guidance levels are given below based on recent analyses.

4.1 Carbon Monoxide Levels and Compliance

Table 10 Carbon Monoxide Ambient Standards and Guidelines

Organisation	Exposure Criteria			
	15 minute	30 minute	1 hour	8 hour
World Health Organisation Guidelines	100 mg/m³ (87.3 ppm)	60 mg/m³ (52.4 ppm)	30 mg/m³ (26.2 ppm)	10 mg/m³ (8.7 ppm)
USA Primary Health Effect NAAQS			40 mg/m³ (not to be exceeded more than once/year)	10 mg/m³ (not to be exceeded more than once/year)

Short-term WHO Guidance

- According to available data[16], the 1-hour, 30-minute or 15-minute WHO guidelines have not been exceeded from 1980–1992. 1973–1979 UK data using a non-dispersive infra-red method have been discounted because of known interference with the measured levels of CO from CO_2.
- Maximum hourly averages for CO (30 μg/m^3 or 25 ppm) have, however, been approached on occasion in urban UK conditions, (24.8 ppm in 1989 in London). These high levels of CO were generally associated with periods of winter stagnation and are most likely to occur in urban environments where the contribution of road transport to CO ambient levels is greatest.
- Hourly guidance levels are likely to be exceeded for non-kerbside urban locations with annual average CO concentrations exceeding approximately 2.5 ppm.

8-Hour WHO Guidance

UK analysis has shown that the 8-hour WHO standard is more stringent than the shorter 1-hour guidelines and, based on log-normal analyses, the 8-hour standard is frequently exceeded in the UK at urban and kerbside locations.

- 45% of urban site measurements exceeded the 8-hour WHO guideline.
- 71% of kerbside site years since 1980 have exceeded the WHO guideline.
- All levels greater than the 8-hour guidance at urban sites have occurred since 1988–1989.
- All levels greater than the 8-hour guidance are associated with stable and cold winter conditions.
- Where annual average CO concentrations exceed approximately 1.25 ppm, it can be expected that the 8-hour guidance may be exceeded.

---

16 M L Williams, 1988, *An Assessment of the UK Position with Respect to the 1987 WHO Guidelines*, WSL Report LR650(AP), Warren Spring Laboratory, Stevenage.

4.2 Nitrogen Dioxide Levels and Compliance

Table 11 NO$_2$ Ambient Air Quality Standards

Organisation	50th percentile of 1 hr means (1 year)	98th percentile of 1 hr means (1 year)	24-hour average	1-hour average	Annual Arithmetic Mean
EC Directive Limit Value		105 (200 µg/m³)			
Guide Value	26 (50 µg/m³)	71 (135 µg/m³)			
World Health Organisation			78 (150 µg/m³)	209 (400 µg/m³)	
USA Primary Health Effect NAAQS					53 (100 µg/m³)

Criteria (ppb)

Table 12 NO$_2$ National Air Quality Public Health Indices

	NO$_2$ Hourly Average Concentration (ppb)			
UK Air Quality Information Criteria	'Very Good' < 50	'Good' 50–99	'Poor' 100–299	'Very Poor' ≥ 300
	NO$_2$ Hourly Average Concentration (ppb)			
US Pollutant Standard Indices	'Hazardous' 2000–1200	'Very Unhealthful' 1200–600	'Unhealthful' < 600	

EC Directive Limit and Guide Values

- The EC **limit** value of 105 ppb has been exceeded 6 times in measurements taken in London, primarily at roadside monitoring stations from the period of 1986–1991. This amounts to 14% of all eligible measurements.
- Limit values are most likely to be exceeded only within close proximity to roads (less than 10 metres distance).
- The EC **guide** value of 71 ppb (at the 98th percentile) is exceeded frequently in London and other UK cities such as Manchester and Glasgow. 23 of 42 measurements (55%) did not meet the health based guide value.

- The EC guide value of 26 ppb (50th percentile) is also frequently violated. Again, 23 of 42 measurements taken over the period of 1986–1991 violated the guide value.
- Correlating monthly NO_2 concentrations to hourly averages, 107 of the 316 diffusion tube survey sites operating in 1991 exceeded the EC 50th percentile guidance value.
- UK populations exposed to NO_2 at levels greater than the EC 98th percentile guidance value increased from 3.8 million in 1986 to 19 million in 1991[17].

WHO Guidance Levels

- Both WHO hourly and daily guidance values are commonly exceeded in the UK.
- The hourly guidance value was exceeded 66 times at three sites in London during 1991 – including an extremely high NO_2 episode from December 12–15, 1991.
- The hourly value was exceeded 118 times from 1988–1991 at nine monitoring sites.
- The daily value was exceeded 101 times from 1988–1991 at the same nine monitoring sites.

---

17 op cit note 12.

4.3 Particulate Matter Levels and Compliance

Table 13 Particulate Matter Ambient Air Quality Standards

Organisation	(μg/m³) 24 hour	Criteria (μg/m³) 1 year	98th percentage of daily means
WHO Guidelines			
Sulphur Dioxide			
Black Smoke	125	50	
TSP	125	50	
Thoracic Particles	120		
	70		
EC			
Limit (BS Method)	111	68	213 (not to be exceeded more than 3 consecutive days)
US			
PM-10	150 (1 day or less)	50	
TSP (no longer in effect)	260	75	

Table 14 Particulate Matter Public Health Indices

Organisation	24 hour	Criteria 8 hour	1 hour
EC Directive			
Health (proposed)		55	
WHO		50–60	76–100
USA			
Health			120[1]

1 Maximum daily 1-hour average not to be exceeded more than one hour per year.

EC Limit and Guide Values

- Sulphur and smoke EC combined limit values were exceeded at 13 of the 166 EC monitoring sites in 1983/84. This has been reduced to three in 1990/91.

WHO Guidance Values

The WHO have set guidelines for sulphur dioxide alone and for SO_2 and smoke (particulates) combined.

WHO SO_2 Value

- The few hourly-averaged SO_2 level measurements indicate that the maximum hourly averages at two sites in London exceeded the WHO guideline in 1983–1984 by a wide margin.
- Based on a log-normal extrapolation of hourly averages to annual averages, it appears that the hourly WHO guidance value is exceeded on a wide basis across the UK. Exceedence would be indicated if annual averages were in the 18–35 µg/m^3 range. The average for all of the UK (approximately 400 sites) was 40 µg/m^3 in 1983/4 and 38 µg/m^3 in 1984/5.
- The WHO 10-minute value is more difficult to interpret. Extrapolations indicate that it is more stringent than the 1-hour guidance value and therefore it can be assumed that the UK would have significant levels of exceedence with this guide as well as the 1-hour guide.

WHO SO_2 and PM Value

- As measured by the UK Black Smoke Method, WHO SO_2 and smoke combined guide values were exceeded by 34 sites in a sample of 1985–86 data. The majority of these exceedences are thought to be due to domestic coal burning and not to transportation.

PM10 and the US Standard

- Based on new PM10 measurements from the UK Enhanced Urban Monitoring Network, the US annual arithmetic mean concentration of 50 µg/m^3 has not been exceeded. No comparable information was available to evaluate UK levels of PM10 and the US 24-hour standard of 150µg/m^3.

Long Range Transboundary Air Pollution Convention SO_2 Protocol

Under the 1979 Geneva Convention on long range transboundary pollution, a SO_2 Protocol was agreed in 1985 requiring signatory countries to reduce their SO_2 emissions by a uniform 30% from 1980 levels by the year 1993. Although the UK was one of five countries which did not ratify the Protocol, emission levels have dropped to the required amount due to other regulations, an economic recession, a decline in SO_2 content of fuels and some emission control on existing power plants.

Under a new target for the SO_2 Protocol, presently being discussed, the UK will be liable for more severe SO_2 restrictions. Based on a critical loads approach, the UK will be liable for an 89% reduction in its SO_2 emissions from 1980 levels. This is much more than it will be able to achieve by complying with the EC large combustion plant directive alone.

Depending on the outcome of these negotiations, the UK, under any circumstance, will be subject to increased pressure to reduce SO_2 emission levels. Although SO_2 from transport is a small contributor, it can be expected that any measures that would reduce ambient SO_2 levels will be under consideration. Very low sulphur fuel oils is one such option that could affect electric generating stations, shipping, commercial heating and road transport.

In addition to the SO_2 Protocol, the UK has ratified the 1988 NO_x Protocol requiring national NO_x emissions to be reduced to 1987 levels by no later than 1994. Even with occasional use of low-NO_x burner technology at power plants and lower passenger car NO_x emissions, meeting the NO_x Protocol may prove difficult.

4.4 Ozone Levels and Compliance

Table 15 National Air Quality Public Health Indices O_3

UK Air Quality Information Criteria	\multicolumn{4}{c}{Hourly Average Concentration (ppb)}
	'Very Good' < 50	'Good' 50–89	'Poor' 90–179	'Very Poor' ≥ 180

US Pollutant Standard Indices	\multicolumn{5}{c}{Hourly Average Concentration (ppb)}
	'Hazardous' 600–400	'Very Unhealthful' 399–200	'Unhealthful' 199–120	'Moderate' 119–60	'Good' 59–0

EC Guidance

In October 1992, the EC published a new Directive on tropospheric ozone, specifying pollutant standard index thresholds and certain minimum monitoring requirements verified by the UV absorption method. The four threshold levels and corresponding advice are as follows:

Health Protection	110 µg/m^3	(8-hour mean)
Vegetation Protection	200 µg/m^3	(1-hour mean)
	65 µg/m^3	(24-hour mean)
Population Information	180 µg/m^3	(1-hour mean)
Population Warning	360 µg/m^3	(1-hour mean)

Public information systems must be activated if the 180 and 360 µg/m^3 levels are exceeded. The EC did not set an ambient ozone standard due to the perceived difficulty of Member-state's complying with one that was established at a level that would take account of health concerns. An ambient ozone standard could be developed in the future.

WHO Guidance Values

- Rural UK ozone levels are higher than urban levels. This is due to the atmospheric chemical transformations that occur over time as ozone pre-cursors move out of urban areas.
- The 80 ppb hourly WHO average is exceeded frequently – up to 18 days and 100 hours per year.
- In urban areas, the upper value 100 ppb WHO average guidance value is exceeded on an infrequent basis. Rural and suburban areas have frequent exceedences up to 10–50 hours in a year.
- Preliminary analysis of the 8-hour average concentrations indicates that the upper end of the range (120 µg/m^3) will be exceeded at most rural and suburban monitoring sites but less frequently at urban ones.
- Based on the UK ambient ozone public health indices, ozone levels classified as "poor" at several urban and rural sites are shown in the figure below.

Figure 19: Hourly ozone concentrations classed as "poor" at selected UK sites, 1990

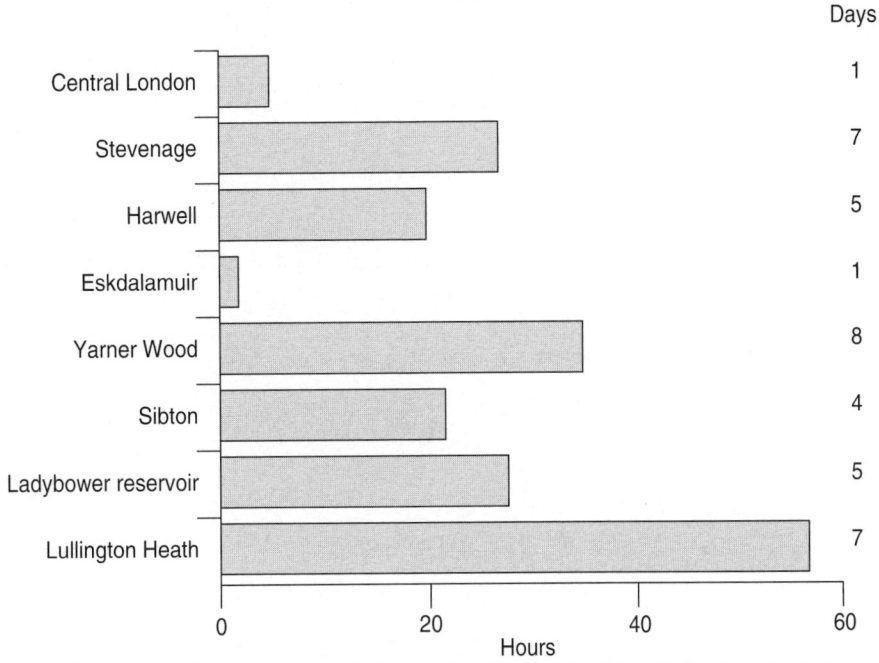

Site	Hours	Days
Central London	~4	1
Stevenage	~27	7
Harwell	~20	5
Eskdalemuir	~2	1
Yarner Wood	~35	8
Sibton	~22	4
Ladybower reservoir	~27	5
Lullington Heath	~57	7

Note: Eskdalemuir (Scottish Borders), Yarner Wood (Devon), Sibton (Suffolk), Ladybower reservoir (Derbyshire), Lullington Heath (East Sussex)

Source: The UK Environment, Department of Environment, HMSO

4.5 CO$_2$ Emission Levels and Progress towards Targets

UK CO$_2$ Stabilisation Goal

The UK has signed the United Nations Convention on Climate Change. Together with other countries, the UK adopted a target of reducing and stabilizing CO$_2$ emissions at the 1990 level by the year 2000. A recent analysis indicates that as a whole, the EC may miss its CO$_2$ reduction target by 4%. The UK, however, has the potential to reduce its projected emissions in the year 2000 to near the target level or indeed below it. Proposals under discussion indicate many areas where emission reductions can be gained. In transport, for example, a 10% rise in real fuel prices could save up to one million tonnes of carbon emissions per year by 2000. Using central estimates of energy use and carbon emissions, this represents 10% of the approximate 10 million tonnes of reduction that are required under the convention target[18].

---

18 Department of the Environment, 1992, *Climate Change: Our National Programme for CO$_2$ Emissions, A Discussion Document*, London.

5 Global Air Pollution and Road Transport

In addition to the local and regional air pollution problems caused by road transport, (which are experienced globally), there are two truly global atmospheric pollution issues to which road transport contributes – global warming and stratospheric ozone. Road transport is currently a major contributor to greenhouse gas emissions and it is likely to be an even greater contributor in the future. Road transport is not the most significant contributor to stratospheric ozone depletion, but advances in ozone depletion policies have targeted the motor vehicle industry as well as others.

Recent policy developments concerning these two global issues are reviewed below. In the case of CO_2 emissions and transport, the review does not cover the various technical or planning measures that are available to reduce such emissions. Rather, it provides a review of some of the CO_2 reduction targets being set in various countries and some of the considerations that are necessary when incorporating CO_2 reductions into an existing array of vehicle emission control policies. Because of the difficulty of setting CO_2 vehicle emission standards, the reduction of CO_2 from road transport is an excellent area to apply market and fiscal incentives to try to achieve efficiency improvements.

5.1 Global Warming, CO_2 Emissions and Transport

Of the gases potentially responsible for global climate warming, road transport CO_2 emissions have received more attention than other potential climate change pollutants[19].

5.1.1 Road Transport CO_2 in the UK and EC

In 1990, the UK contributed 2.7% of the estimated world global emissions of CO_2 – or 160 million tonnes of a total of 6,000 million tonnes. Total transport accounted for 33 million tonnes of carbon. Passenger cars account for 13% of the UK total[20].

---

19 CO emissions, for example, can lead indirectly to increases in concentrations of methane in the atmosphere. The various CFCs and HCFCs used in the production and operation of road vehicles are greenhouse gases in addition to stratospheric ozone depleters.
20 op cit note 18.

Road transport in the EC accounts for more than 20% of CO_2 emissions and over 60% is due to passenger cars[21].

5.1.2 Transport CO_2 Trends

Transportation CO_2 emissions amounted to 20 mtC in 1970 (11% of total UK emissions), 33mtC in 1990 (21%) and are estimated to be around 55mtC (25%) in 2020. Based on an estimated increase in road traffic demand of between 69 and 113% by the year 2025, CO_2 emissions from transport will not only grow in absolute terms, but will also increase in relative terms to other sources.

Recent research conducted for World Wide Fund for Nature UK estimates the CO_2 contribution of passenger cars into the future, and also quantifies the emissions of nitrous oxide and methane, two other greenhouse gases[22]. It is important to incorporate all greenhouse gaseous pollutants into a future control strategy because of the differential 'global warming potential' of various chemicals. Methane, for example, has an estimated global warming potential approximately 50 times greater than CO_2 over a 20 year time projection.

According to a base case scenario, the report estimates the following changes in pollutant emissions from passenger cars in the UK between 1990 and 2025.

Table 16 UK Passenger Car Greenhouse Gas Emissions 1990–2025

Gas	Year	Emissions (tonnes)	Percentage change (1990–2025)
CO_2	1990	72,404	
	2025	116,024	+60.2
CO	1990	6,328	
	2025	2,076	−67.2
CH_4	1990	20	
	2025	11	−41.5
NMVOC	1990	783	
	2025	146	−81.4
NO_x	1990	1,055	
	2025	271	−74.3
N_2O	1990	2	
	2025	25	+1,336

Source: Wade, Holman and Fergusson, Earth Resources Research, 1993

21 European Commission, 1991, *Motor Vehicles' Fuel consumption and CO_2 emissions*.

22 J Wade, C Holman, M Fergusson, 1993, *Current and Projected Global Warming Potential of Passenger Cars in the UK: A Report for World Wide Fund for Nature UK*, Earth Resources Research, London.

Notable are the increases in CO_2 and N_2O. CO_2 increases are a result of increases in demand for transport in advance of technical efficiency improvements or curbs in demand brought about by proposed fuel tax increases. Increases in N_2O are a result of estimated increased N_2O emissions from three-way catalyst controlled vehicles compared to non-catalyst equipped cars (from 5 to 10 times higher in the former).

When these results are interpreted from the perspective of the Climate Change Convention, CO_2 and N_2O emission projections from this sector do not meet the goal of stabilization of 1990 emission levels by the year 2000. The base-case scenario estimates passenger car CO_2 emissions in the year 2000 increasing by about 19% over the 1990 level. Under a low traffic growth forecast, the 2000 level is 14% greater than the 1990 level.

The report also translates these *emission* projections into *global warming potential* projections. From this perspective, despite the increases in CO_2 and N_2O, the total passenger car global warming emissions potential is only 3% greater in 2000, than in 1990 and is 1% less in 2003, than the 1990 level. This reduced global warming impact (due to reductions in conventional pollutants such as NO_x, CO, and non-methane hydrocarbons which all have a global warming potential) is maintained until the year 2015. At that point, growth in overall transport demand and corresponding CO_2 emissions outweigh the benefit of reductions gained through application of the catalyst to the passenger car fleet.

Other important conclusions from the study include:

- In the long term, the total warming impact of passenger cars will grow unless technological improvement is substantial or demand is managed, limited and shifted to alternative modes of transport.
- This conclusion holds true even if the diesel passenger car market increases from an estimated 20% of the total new car sales in 2025 to 40%. Although emissions from diesel passenger cars have a future global warming potential of approximately 20% less than the average petrol car, a substantial impact on total fleet greenhouse emissions is difficult to achieve. Any increases in the diesel car proportion of the fleet must also be evaluated from a public health and ecological perspective.

The increased nitrous oxide emissions resulting from application of the 3-way catalyst to cars needs technical attention. In addition the emission factors used to estimate N_2O emissions need further attention.

5.1.3 CO_2 Transport Reduction Proposals

Germany

Germany has proposed reducing CO_2 emissions from road transport by approximately 40% between 1993 and 2005 according to a vehicle weight class scheme.

Germany's proposal to develop a revenue-neutral scheme, which reduces tax on cars with lower NO_x, CO, HC, PM and noise emissions, was also going to incorporate CO_2 emissions (thus fuel economy). Recent developments indicate that CO_2 will not be part of the initial programme proposal however[23].

Netherlands

The Dutch have considered a proposal to reduce overall CO_2 by 35% by the year 2010–40% from heavy vehicles, and 30% from lighter ones. This is part of their wider National Environmental Policy plan which identifies a road transport CO_2 emission goal of stabilization by 1994, 3–5% reduction by the year 2000, and 10–20% by the year 2010[24].

Japan

Government advisers recommend an improvement of auto fuel efficiency by 8.5% from current levels by the year 2000. The new standards will result in a minimum average mileage of 12.3 km/litre (about 30 miles per gallon) from the current 11.34 kilometres. Also contained in the report was a recommended requirement for auto manufacturers to use more aluminium, instead of plastics, to reduce vehicle weights and to promote the research and development of lean burn engines[25].

UK

In the Chancellor's March '93 budget statement, the UK government announced an intention to raise road fuel duties, on average, by at least 3% a

---

[23] M P Walsh, 1992, *Car Lines, International Regulatory Developments: The Year in Review*, volume 9, number 6, Arlington, VA.
[24] Concawe, 1992, *Motor vehicle emission regulations and fuel specifications – 1992 Update*, Report no. 2/92.
[25] op cit note 24.

year in real terms in future years. This is in addition to recently announced 10% increase in road fuel tax for '93–'94. CO_2 reduction is an explicit goal of this taxation programme. The Government has stated that a 30% increase in real fuel prices by the year 2000, could result in a 2 million tonne carbon reduction from the transport sector[26].

The Society of Motor Manufacturers and Traders (SMMT) have adopted an ACEA proposal and set a voluntary fuel efficiency improvement target of 10% by 2005. This is a sales weighted average proposal that would be achieved both by improvements in vehicle technology and shifts in consumer demand to more efficient vehicles.

The UK non-governmental environmental organization, Friends of the Earth, has proposed improvements in fuel economy ranging from 45% for vehicles of 800 kg in weight or less, 50% for vehicles 800–1,100kg in weight and 55% for vehicles from 1100–1,500 kg/gvw. These efficiency standards should be written in terms of CO_2 limits (gram/km), which will allow for the evaluation of alternative fuels from a CO_2 emission perspective. Thus, fuels with a lower CO_2 emitting potential would have an advantage of meeting the reduction targets over petrol, and fuels with a greater CO_2 emission potential would find it more difficult to meet the target[27].

5.1.4 Transport Fuel Efficiency Standards

One of the main regulatory forces in achieving reduced CO_2 from road transport is the existing fuel efficiency targets and standards operating in various countries. The existing standards in the US and Japan and the trends that they will be following in the future will be an enormously important factor in the European transport sector's efforts to improve efficiency and reduce emissions. The other major factor affecting efficiency improvements in the European transport market are relative fuel price increases, and the subsequent shift in demand for more fuel efficient vehicles.

In the US, there have been recent legislative proposals designed to improve the passenger car fleet efficiency standard by approximately 40% by the year 2001. This would raise the fleet standard from the current 27.5 miles per gallon (11.5 km/litre) to 40 miles per gallon (16.7 km/litre). These proposals were not adopted in the 1991–1992 legislative year, but are likely to be re-introduced in the 1993–1994 Congressional session.

---

26 op cit note 15.

27 Friends of the Earth, 1991, *Climate Change and Cars in the EC: Technological Standards to Reduce Carbon Dioxide Emissions from Cars in the EC*, briefing sheet, London.

In addition to setting a standard for domestic and foreign manufacturers, the current US programme carries incentives for the production of alternative fuelled and non-gasoline vehicles. Vehicles fuelled with ethanol or methanol or a mix of these fuels with petrol receive extra credit in terms of fuel economy.

Implications and Future Directions for the UK

- Further increases in fuel efficiency standards in the US and Japan will create a demand for more fuel efficient vehicles by manufacturers world-wide and will create potential opportunities for increased foreign sales of diesel passenger cars, which are currently a small fraction of the US market. It should be noted however, that any increase in US diesel sales will be difficult. Current and new diesel vehicle emission standards represent a technological barrier.
- The current US system of regulating fuel economy does not take into account the energy used in producing or distributing the fuels burned. Under such circumstances, some fuels which have high energy production penalties (eg electric vehicles) are given more incentive than they may receive in a European efficiency programme, which is more likely to incorporate a life-cycle analysis that acknowledges a fuller energy cycle.

5.1.5 Encouraging More Fuel Efficient Cars: The Impact of Fleet Efficiency Standards and Fuel Price

There has been a long-standing debate about whether vehicle fuel efficiency standards or increases in fuel price are responsible for the noticeable increase in average fuel economy of passenger cars. In the US, the average fuel economy of new passenger cars increased from 14 mile/gallon (5.9 km/litre) in 1974 to 28 miles/gallon (11.7 km/litre) in 1988.

Two major events could be responsible for the change in technical efficiency: fuel prices doubled over the period (yet in constant 1988$, the 1989 price of $0.96 per gallon was actually below the 1975 price); and CAFE (Corporate Average Fuel Economy) standards for cars were set (18 mpg by 1978 and 27.5 mpg by 1985).

In a statistical analysis of the power of both price and standards, it was definitively shown that for the big 3 auto manufacturers in the US and some European manufacturers, standards were at least twice as important as increases in fuel prices in producing more fuel efficient cars, and over the long-term, standards may completely have outweighed the forcing function of fuel price[28].

Although this research documents the effect of standards on efficiency improvements in the US, it is difficult to translate the experience to the European and UK markets. Here, due to greater initial petrol prices, it is likely that price increases may affect demand more than they would in low fuel price markets. This is the approach that is being adopted in various countries including the UK as indicated above. In Germany, the lower house of Parliament recently approved an increase of 17% on petrol and 13% on diesel fuel, in most part to support federal budget priorities, but also to foster conservation and investments in efficiency[29].

5.1.6 CO_2 -versus- NO_x Reduction

One of the major dilemmas facing vehicle manufacturers and policy makers is optimizing both the reduction of carbon dioxide and conventional pollutants. The challenge of reducing CO, HC, and NO_x at the same time resulted in the development of the 3-way catalytic converter. Optimizing the control of CO_2 in addition to these pollutants will be even more difficult. The main reason for this is the fuel economy penalty experienced by 3-way catalyst equipped cars.

Many studies have diagnosed this relationship. One indicated that the reduction in the US NO_x standard from 1.0 g/mile to 0.4 g/mile, and the application of a 3-way catalyst equipped car to meet that standard, results in an average fuel efficiency penalty of up to 7%. This penalty is increased by a further 7–8% when the NO_x limit is reduced to 0.2 g/mile[30]. An earlier study isolated a difference of about 4% between catalyst-equipped and non-catalyst equipped cars. Annual trends in overall registration weighted fuel consumption by new passenger cars does not show a decrease in overall fuel efficiency, however. Better engine performance management may have offset the potential fuel economy disbenefit associated with catalyst-equipped cars[31].

Another study of the relationship between emission control systems, fuel efficiency effects and reductions of conventional pollutants concluded[32]:

28 D L Greene, 1990, "CAFE or Price?: An Analysis of the Effects of Federal Fuel Economy Regulations and Gasoline Price on New Car MPG, 1978–1989", 1990, *The Energy Journal*, 11:37–57.
29 ENDS Report 221, 1993, page 39.
30 H P Lenz, 1992, *NO$_x$ vs. CO$_2$: Literature Study on Emissions, Air Quality and Effects of Vehicle Exhaust Components*, Institute for Internal Combustion Engines and Automotive Engineering, Vienna Technical University.
31 R O'Reilly, 1993, UK Department of Transportation, comments on draft report.
32 J M Dunne, 1990, *A Comparison of various emission control technology cars and their influence on exhaust emissions and fuel economy*, Warren Spring Laboratory, LR 770 (AP), Stevenage.

- Averaged across a range of vehicle speeds, a fuel economy penalty of 9% was experienced for 3-way catalyst equipped cars conforming to the EC 15.04 regulation during road tests. Dynamometer tests indicated a greater penalty of around 20%.
- CO_2 emissions increased by 24% for low-speed road-tested vehicles and 8.5% for high-speed vehicles.

This dilemma is being addressed technically in a number of ways. The development of zeolite supported catalysts for use in lean-burn engine designs is one. Normally, under lean-burn efficient operating conditions, the performance of a 3-way catalyst is maintained for the control of carbon monoxide and hydrocarbons but is reduced for the control of NO_x. This problem is solved in stationary sources with the application of SCR (selective catalytic reduction), using additives such as ammonia. This is not possible for transportation applications. For mobile sources, catalysts that utilise reduction agents in the exhaust gas stream itself are one potential technical solution. Although progress is being made, much work remains before practical application can proceed[33].

Implications and Future Directions for the UK

- Adding fuel efficiency or CO_2 emission standards to the existing array of vehicle emission standards is complicated. It must be done in full recognition of the difficulty of improving both efficiency and reducing conventional pollutants.

5.1.7 CO_2 Life cycle emission analysis

Optimal public policy aimed at the reduction of any air pollutant should take into account the total emission portfolio offered by the use of various fuels and various modes of transport. For instance, CO_2 emissions from a car's operation should be considered together with the CO_2 emitted from fuel production, car production, and car disposal. One such analysis indicated that average vehicle operation contributed to over two-thirds of CO_2 emissions on a gram/km basis life-cycle analysis. Thus, vehicle operation regulations should continue to be a focal point of policy. But as vehicle emissions are controlled, the fuel, vehicle production and post-operation phase emissions must be taken into account in order to further refine policy[34].

---

33 T J Truex, R A Searles, D C Sun, 1992, "Catalysts for Nitrogen Oxides Control under Lean Burn Conditions: The opportunity for new technology to complement platinum group metal autocatalysts," *Platinum Metals Review*, 36:2–11.

34 D Martin and L Michaelis, 1992, *Road Transport and the Environment: Policy, technology and market forces*, Financial Times Business Information Report, London.

5.2 Stratospheric Ozone Depletion and Road Transport

Another adverse environmental effect of the road transport sector is its contribution to increased atmospheric concentrations of CFCs and HCFCs and the resultant depletion of the stratospheric ozone layer. This section briefly describes the state of research concerning the depletion of the ozone layer, the road transport sector's use of ozone-depleting substances, and current and future policy developments concerning such use.

The depletion of the stratospheric ozone layer, more than any other environmental issue, documents the necessity for foresight.

Although CFCs were discovered in the latter part of the nineteenth century, widespread use did not occur until the 1940s and 50s. If the chemicals had been widely used before the 1950s it is very likely that the Earth's protective ozone layer would have been destroyed before it had even been discovered.

5.2.1 Ozone Depletion

With the increase in atmospheric concentrations of man-made gases (CFCs and HCFCs), there has been a significant depletion in the level of stratospheric ozone. The effect is most dramatic over Antarctica where an 'ozone-hole' now forms in the autumn of every year. Recent studies indicate that:

- Stratospheric ozone levels of northern latitudes are falling. A 6-8 % decline has been observed in the early months of the year[35].
- The loss of ozone in the northern mid-latitudes (between 1979–1990) is more than 8% in early spring each year[36].

The decline in Antarctic ozone levels is illustrated in the Figure 20 below.

5.2.2 Effects of Ozone Depletion

'The ozone layer in the stratosphere helps prevent very harmful short wavelength (less than 280 nm) ultraviolet-C radiation from the sun penetrating to the surface of the earth, and substantially reduces the amount of harmful ultraviolet-B (wavelength 280–320 nm) radiation reaching the earth's surface. Ultraviolet-B can cause skin cancer and can have harmful effects on plants (including agricultural crops) and marine organisms[37].''

---

[35] Department of the Environment, 1992, *The UK Environment*, editor Alan Brown, HMSO, London.
[36] op cit note 35.
[37] op cit note 35.

Figure 20: Ozone levels in the Antarctic[1], 1957 - 1990

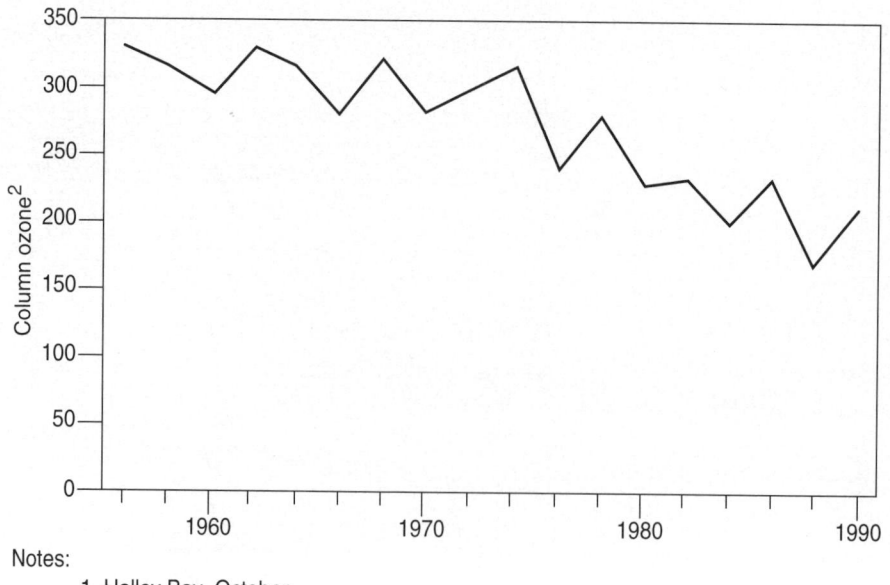

Notes:
1. Halley Bay, October
2. Dobson Units

Source: UKSORG, The UK Environment, Department of the Environment, HMSO

The hazards of stratospheric ozone depletion and ozone-depleting compounds include the harmful effects of increased ultraviolet (UV-B) radiation on human health and the environment, and increased global warming. Examples of some of the effects caused by the depletion include:

- increases in the number of skin cancer cases (it has been estimated that 1% decline in ozone has been linked with a 2% increase in skin cancer incidence);
- increased incidence of cataracts;
- suppression of the immune response system of humans and animals;
- increased tropospheric smog formation; and
- accelerated deterioration of plastics and other materials.

5.2.3 Stratospheric Ozone Depleter Use in the Transport Industry

In the overall UK economy, statistics on the amount of ozone depleters used or emitted by the transport sector are not readily available. Table 17 below outlines recent trends in CFC and halon use in the UK economy as a whole and Table 18 outlines the percentage of these chemicals used for various end purposes.

Table 17 UK Consumption of CFCs and Halons (000 tonnes weighted)

	1986	1989
Total CFCs	62.8	31.3
Total Halons	9.0	10.0
Total Controlled CFCs & Halons	71.8	41.3

Source: DTI CFCs & Halons, 1990

Table 18 UK Consumption of CFCs by Primary Area of Application

	1986 %	1989 %
Aerosols	64	24
Foam Blowing	16	28
Refrigeration & Air Conditioning	11	31
Solvents	9	17

% weighted by ozone depleting potential

Source: DTI CFCs & Halons, 1990

The transport industry uses CFCs and halons during production, manufacturing and use phases. Blown plastic foams for seat cushions and insulation, air conditioning refrigerants, and solvents have been the traditional uses.

Blown Plastic Foams

Polyurethane foam is used as insulation for refrigeration applications. In the transport industry it is used in the construction of steering wheel covers, dashboards, headrests and seat cushions (where demand has peaked), and for sound and vibration insulation, where demand will increase in the future due to continued efforts to control noise by encapsulation. CFCs 11 and 12 have dominated in the production of blown foams.

Consumption of polyurethane foam in the UK vehicle industry grew from over 12.5 million tons in 1984 to almost 16 million tons in 1987[38]. Recently, however, the trend in the industry is to move away from its use in both hard and soft form.

---

38 Coopers and Lybrand Deloitte in association with Mott Macdonald and C S Todd & Associates, 1990, *CFCs and Halons: Alternatives and the Scope for Recovery for Recycling and Destruction*, HSMO, London.

Refrigeration and Air Conditioning

The transport sector uses CFCs and halons in transportation refrigeration and air conditioning. As a whole, the refrigeration and air conditioning industry used about 10–12% of the total CFC consumption in the UK in 1986. Transport refrigeration represents the smallest use of CFCs in the refrigeration and air conditioning sector (about 1%). In 1988, the Department of Transport estimated that the refrigerated truck fleet in the UK numbered approximately 34,000 vehicles. The report estimated that transport refrigeration uses nearly 60 tonnes of CFC per annum in the UK, with approximately 75% used by trucks.

Lorries primarily use CFC-12 as a refrigerant. Alternatives to CFC-12 (and HCFC-22) include halogenated chemical alternatives such as HFC-134a and non-halogenated chemical alternatives such as ammonia. Concerns over the flammability and toxicity of ammonia have prevented its application in mobile sources to date, but it is increasingly being used in stationary source refrigeration, such as off-site central cooling facilities for grocery stores.

Mobile air conditioning is another potential end use of ozone depleters in the motor vehicle industry. Information about the number of air conditioned vehicles and the amount of CFC chemicals used in such air conditioners was not available. However, mobile air conditioning has received a lot of policy attention in the US.

5.2.4 US Policy Developments

In addition to strengthening some of the provisions of the Montreal Protocol, the Clean Air Act of 1990 required the EPA to implement a national recycling and emission reduction program. This programme will include regulations regarding the use and disposal of class I and II substances during the service, repair, or disposal of appliances and industrial process refrigeration.

The EPA will also promulgate regulations by early 1992 for the servicing of motor vehicle air conditioners, as well as regulations requiring product labelling and the banning certain non-essential products.

Motor Vehicle Air Conditioners

Mobile air conditioners in the US represent a large portion of the residual demand for CFC use. Residual demand is defined as CFC-based equipment and end uses that will have demand for the product after any phase-out.

Because mobile air conditioners represent around 21% of the residual demand for CFCs, it was important to implement a programme for recovery and recycling to assist the continued servicing of these CFC-based uses[39]. Residual demand for CFC-based end uses in the UK is concentrated in the commercial refrigeration, sterilizers, chillers, and not in mobile air conditioners.

EPA estimates that 50% to 94% of the emissions from mobile air conditioning equipment and appliances occur during servicing and disposal; thus CFC recycling and recovery programmes were deemed crucial. Through the proposed programmes described below, EPA expects the mobile source air conditioner programme to decrease ozone depleter releases from 41,000 metric tons to 15,000 tons per year[40].

The US Federal Programme

From 1992 (or 1993 for service stations servicing fewer than 100 motor vehicles air conditioning units annually), stations servicing or repairing such units were required to use approved refrigerant recycling equipment. Servicers were also required to be trained and certified to use such equipment. In 1992, it became illegal to sell or distribute motor vehicle refrigerant in containers under 20 pounds in weight. This effectively eliminated the 'home-repair' market for mobile air conditioners, where leaky conditioners were often topped-off with refrigerant by car owners.

In addition, according to the proposed regulation, the EPA was required to establish a technician and equipment certification programme that would ensure that:

- All recycling and recovery equipment sold would be capable of minimizing emissions;
- technicians would be properly trained in methods to cut emissions;
- Reclaimed refrigerants being sold would be of high quality to avoid problems;
- In addition, sales of refrigerants would be restricted only to qualified technicians.

The California Programme

In addition to the US regulatory requirements, California has enacted a bill proposing to phase-out CFC refrigerants in new motor vehicles sold in

---

39 US Environmental Protection Agency, 1989, *Costs and Benefits of Phasing Out Production of CFCs and Halons in the United States*, Office of Air and Radiation, Washington DC.
40 op cit note 6.

California. The phase-out programme requires that not more than 90% of vehicles sold, supplied or offered for sale in California in 1993 may use CFC-11 or CFC-12 in mobile air conditioners. This will be reduced to 75% in 1994, and completely banned from January 1995.

Industrial Effects

As a result of international and federal controls on ozone depleters, increased opportunities will occur for the development of CFC substitutes, the manufacture, design and construction of CFC recovery and recycling equipment, including leak detection equipment, and the development and production of non-CFC containing product substitutes.

It is difficult to isolate the increased sales revenues that will accrue due to the motor vehicle industry changes. The overall effect on US industry and economy is expected to be around $300 million in increased sales revenues (excluding sales of substitute chemicals – in 1990$) on average during 1992–1995, resulting in a cumulative revenue increase of about $2 billion over the 1992–2000 period. Revenues for environmental service companies are expected to grow significantly[41]. These potential revenue increases are given for illustrative purposes only. All too often the benefit side of environmental regulation remains unquantified. Obviously, these benefits would need to be weighed against the potential cost of the regulations.

The motor vehicle manufacturing industry is responding to the challenge by eliminating CFC usage in auto manufacturing. In Japan, all auto makers plan to stop using chloroflorohydrocarbons and methyl chloroform by the end of 1994 or earlier. Nissan has stopped using CFCs in manufacturing and will stop using them in air conditioners by the end of 1994.

Recent Global Regulatory Development

Table 19 outlines the recently agreed updates to the original Montreal Protocol –the so-called Copenhagen Deadlines. The EC has agreed to adopt these global deadlines and has set more stringent deadlines for the phase-out of CFCs and carbon tetrachloride. Further proposals for phase-out of methyl bromide, HCFCs, and hydrobromofluorocarbons may be developed in 1993 at the EC level.

---

41 Smith Barney & ICF Resources Inc, 1992, "Business Opportunities of the New Clean Air Act", prepared for US Environmental Protection Agency, Washington.

Table 19 Stratospheric Ozone Depleter Phase-Out Deadlines

Global Deadlines

Year	CFCs	Halons	Methyl Chloroform	Carbon Tetrachloride	Methyl Bromide	HCFCs
1994	75% cut	100% cut	50% cut			
1995				85% cut	cap begins	
1996	100% cut		100% cut	100% cut		cap begins
2004						35% cut
2010						65% cut
2015						90% cut
2020						99.5% cut
2030						100% cut

EC Deadlines

Year	CFCs	Halons	Methyl Chloroform	Carbon Tetrachloride	Methyl Bromide	HCFCs
1994	85% cut	100% cut	50% cut	85% cut	proposals expected by March/April 1993	proposals expected by March/April 1993
1995	100% cut			100% cut		
1996			100% cut			

Notes: Most dates refer to 1 January. Base year for halons and most CFCs is 1986 (those CFCs first controlled under the 1990 London amendments use 1989 as the base year). Base year for methyl chloroform, carbon tetrachloride, and HCFCs is 1989. Methyl bromide cap is set at 1991 levels. Hydrobromofluorocarbons (HBFCs) are phased out fully in 1996.

Source: Environment Watch, Special Edition, 1993

In the area of ozone depleter phase-out and production, the ban of CFCs for refrigeration equipment will most affect the motor vehicle industry. Belgium has banned CFCs 11, 12, 113, 114, 115 since 1991; Canada is considering mandatory recovery and recycling programmes for CFCs in use and Germany is phasing out use of CFCs for mobile truck air conditioners from January 1994 and HCFC for the same purpose by January 2000. Dutch use of CFCs and Halons is expected to be zero for purposes of new refrigeration by the year 1995.

The UK does not have any specific mobile source regulations and will adhere to the EC phase-out schedule.

Implications and Future Directions for the UK

- It is likely that as CFC use is phased out and the phase-out pressure on substitutes known to contribute to ozone depletion becomes greater, the residual amount of CFCs in various end-uses will become likely targets for recovery and recycling programmes.
- Given that the UK auto fleet is relatively un-air conditioned, it is unlikely that a recover and recycle programme will be developed in the UK.
- Regulations covering the servicing of such existing equipment, especially in the case of the refrigerated freight fleet should be anticipated, however, as a necessary development.
- Research is needed to identify the need for and potential size of CFC and halon 'banks', the sources of such 'banks' and the establishment of programs to establish, service and operate these banks through the development of CFC recovery and recycle programmes.

6 The Health Effects of Road Transport Air Pollution: Current and Future Policy Developments

6.1 Introduction

To date, motor vehicle emission control policies have been implemented primarily for the protection of human health and secondarily for the protection of ecological health. Motor vehicle emission control policies have concentrated on the precursors of tropospheric ozone or smog (HC and NO_x) and on commonly known toxic emissions, such as lead, carbon monoxide, and particulate matter. Initial attention on hydrocarbon emissions, because of their contribution to the formation of smog, has been followed by attention to NO_x emissions for their contribution to smog formation and for the harmful effects of NO_2 itself. HC, NO_x, and CO control strategies have become progressively more stringent over time. Pollutants of recent concern include specific air toxics (addressed primarily for their carcinogenic potential) and pollutants contributing to global air pollution problems, such as CO_2 and CFCs. In addition, as motor vehicle fuels other than diesel and petrol gain favour, their emission properties are becoming increasingly scrutinized for any adverse health effect potential.

The origin of tailpipe emission standards in California and at the federal US level was designed to help alleviate local and regional ambient air pollution problems (primarily smog) and eventually to help attain ambient air quality standards. These standards for particulate matter, sulphur oxides, carbon monoxide, nitrogen dioxide, ozone and lead were initially determined through a review and analysis of the health effects research of these pollutants. National ambient air quality standards (NAAQS), with averaging periods varying from one hour to one year, were established. The levels set in the United States were not 'no effects' levels or thresholds, but levels designed to 'protect human health with an adequate margin of safety.'

Ambient air quality standards do not by themselves determine the exact emission standards for motor vehicles, but they are a driving force behind the evolution of such standards. It is therefore necessary to examine how changes in the underlying health information could result in changes and modifications to ambient air quality standards and, in turn, to vehicle emission limitations and fuel specifications. Changes in our understanding of the ill effects of motor vehicle pollution is also a motive force in the evolution of other types of policies, such as traffic planning.

This section of the report will review developments in health effects research of 'conventional' mobile source air pollution (SO_2, NO_2, ozone, particulate matter), and will review in more detail the health effects associated with hazardous air pollutants such as benzene and 1,3-butadiene.

It is important for the reader to understand the purpose of reviewing health research developments in this report. The main purpose is to draw attention to one of the dominant factors shaping road transport pollution control policy today. Policies and standards are already in place because of health concerns. These policy trends are being replicated in countries other than their origin.

Vehicle and fuel technical development is being transformed because of these research developments and concerns. This review does not attempt to evaluate the certainty of the studies cited. Instead, it simply illustrates those recent health-based concerns that are already shaping the road transport market and will likely affect it to a larger degree in the future.

It is important for the UK to be aware of the motive forces behind environmental policy world-wide. With this foresight, researchers and policy makers can determine whether the policy and technical trends portrayed are warranted by the science. If, in their opinion, they are not, then it is imperative that these concerns be voiced in international forum where policies affecting the UK transport and fuels market are being made.

6.2 Developments in 'Conventional' Air Pollution Health Research

In the US and other countries, the health effects of so-called 'conventional' air pollutants such as carbon monoxide, nitrogen oxides and ozone are under continual review. In the US, the purpose of such a review is to ascertain the adequacy of current ambient air quality standards. The standards are being reviewed for the level of protection afforded not just by the ambient concentrations allowed, but also for the period of averaging time used to determine compliance with each standards. In the past few years, research has been expanded to look at the effects of long-term relatively low-level concentrations of pollution. This work and other developments are briefly described below.

NO_2

Recent research indicates that short duration human exposure to NO_2 at levels of 1.5-2 ppm produce increased lung reactivity to various pollutants. In addition, asthmatic individuals show increased bronchial reactivity during short term exposure to NO_2 levels of around 1 ppm. Animal studies indicate that lung defense systems are weakened by long-term, low-level NO_2 exposure and respiratory infections in children exposed to low levels of NO_2 for long periods of time have been reported[42].

This recent research builds on earlier work showing demonstrated effects in asthmatics exposed to levels of NO_2 levels of approximately 300 ppb[43]. This work resulted in the existing WHO guideline (1 hour average) for NO_2 of 210 ppb.

Ozone

Inhalation exposure to ozone at levels below the current US ambient standard is responsible for decreased respiratory function in exercising healthy adults[44]. Long term ozone exposure can lead to chronic lung development and function damage.

The transient effects of ozone appear to be more closely aligned to cumulative daily exposure than to one-hour peak concentrations. Although the effects of long term chronic exposure to O_3 remain poorly defined, recent animal and epidemiological studies suggest that current ambient levels of 12 ppm or 240 $\mu g/m^3$ are enough to cause premature lung aging[45].

SO_2

Ambient concentrations of short term SO_2 concentrations of 0.2–0.3 ppm (an order of magnitude lower than concentrations that affect the general population) are reported to have adverse respiratory effects[46]. Adverse health effects at ambient concentrations lower than the WHO lowest observed effect levels

---

42 L Grant, 1992, "The health effects of criteria air pollutants (nitrogen oxides, carbon monoxide, ozone and sulphur oxides)", US Environmental Protection Agency, HERL, USA, *Report of an International Programme on Chemical Safety, Australia International Workshop on Human Health and Environmental Effects of Motor Vehicle Fuels and Exhaust Emissions, Interim Report*, United Nations Environment Programme.

43 World Health Organization Regional Office for Europe, 1987, *Air Quality Guidelines for Europe*, WHO Regional Publications European Series No. 23, Copenhagen.

44 op cit note 42.

45 M Lippman, 1989, "Health Effects of Ozone: A critical review", *Journal of the Air Pollution Control Association*, 39: 672–695.

46 op cit note 42.

for short and long term average concentrations in 1987, have been documented[47].

Acid Sulfate Aerosols

Acid aerosols are known to have adverse effects on respiratory health. Lung function is altered after acute exposure and sulphuric acid appears to be more damaging than other sulphates. A re-analysis of the London smog excess mortality incidence, indicates that acid aerosols (SO_4) were the component of the smog responsible for the most damage. This re-analysis remains uncertain. Other research implicates acid aerosols in causing lung function changes in children and in being responsible for acute hospital admissions during periods of elevated sulphate levels. Aerosols may act synergistically with ozone to produce elevated respiratory health effects[48].

A review of this material to 1990 concludes that secondarily formed aerosols such as acid sulphate are strongly involved in the adverse effects of the SO_2/particulate complex of chemical pollution[49]. More recent work by Schwartz indicates a strong statistical relationship with elevated acid aerosol and PM-10 levels and mortality and that the relationship is significant at ambient levels of particulate matter well below the health-based US standards. While disagreement remains concerning the nature of the statistical relationship and the models used to construct the analysis, the work is receiving attention from policy makers. Most recently, the US EPA has announced a reevaluation of the health-based air quality criteria on which the existing US PM-10 ambient air quality standard is based.

Implications and Future Directions for the UK

- Most of the epidemiological research (either occupational or community based) has centred on acute health effects of short-term exposure to pollutants such as NO_2, SO_2, and O_3.
- As new research defines the health effects of long-term, lower level exposures, it can be expected that current ambient air pollution standards will be seriously reviewed for the adequacy of protection that they avail.

---

47 I Romieu, 1992, "Epidemiological Studies of the Health Effects of Air Pollution Due to Motor Vehicles", WHO, Mexico, in *Motor Vehicle Air pollution: Public Health Impact and Control Measures*, editors D T Mage, O Zali, WHO, Geneva.

48 op cit note 45.

49 J D Spengler et al, 1990, "Acid air and Health", *Environmental Science and Technology*, 24:946–956.

- In particular, low levels of particulate matter-related acid aerosols increased morbidity and low levels of O_3 and NO_2 increased respiratory health effects among both sensitive populations (asthmatics, exercising adults), and general populations. The relationship between PM and acid aerosols is still not fully understood.
- As air pollution monitoring data in the UK increases, low-level long term exposure research will advance and a review of the adequacy of current standards may result.

6.3 Mobile Source Hazardous Air Pollutants

In addition to more conventional air pollutants (ozone, lead, carbon monoxide, sulphur dioxide, suspended particulate matter and nitrogen dioxide), motor vehicles and motor vehicle fuels are also responsible for the emission of various chlorinated and non-chlorinated hydrocarbons and volatile organic chemicals (VOCs), which have the potential to cause serious health risks to exposed human populations.

So-called 'toxic air pollutants' or 'hazardous air pollutants', such as benzene, 1,3-butadiene, formaldehyde, and particulate matter from engine exhaust, present cancer and/or non-cancer human health risks to exposed populations. These chemicals are emitted due to their presence in motor vehicle fuels as part of the feedstock and are formed secondarily as part of the incomplete combustion process itself.

6.3.1 Motor Vehicle Contribution to Ambient Levels of Air Toxics

The motor vehicle is a major contributing source of the ambient levels of benzene and 1,3-butadiene and other toxic volatile organic chemicals such as toluene. The contribution of road transport to levels of air toxic emissions can also be inferred in a general manner by noting the emission of VOCs from mobile compared to stationary sources. As mentioned above, road transport emitted 41% of the total UK VOC inventory in 1990. Table 20 below notes the contribution of motor vehicles to three hazardous air pollutants in the UK and the US.

Benzene is present in both exhaust and evaporative emissions from mobile sources. Variation in emission levels depends on fuel composition, and absolute levels depend on control technology. Benzene levels in fuel in the US average under 2% and will be limited to 1% in the new reformulated petrol programme. In the UK, benzene content in petrol averaged 3.1% in 1986 and 2.2% in 1987 and is limited to an EC maximum of 5%.

Table 20 Motor Vehicle Contribution to Ambient Air Toxic Concentrations

	US EPA[50]	UK DOE[51]
Benzene	85%	89%
Formaldehyde	33%	–
1,3-Butadiene	94%	100%

Ambient levels of **1,3-butadiene** are also highly dependent on mobile source emissions. 1,3-butadiene emissions are a result of incomplete combustion of the fuel and/or the cracking of higher olefins in higher grade and unleaded fuels. Other ambient sources of this chemical are due to the production of 1,3-butadiene and rubber industry production, a large commercial user of the chemical.

Identifying the contribution of mobile sources to ambient levels of **formaldehyde** is more difficult because formaldehyde is formed as a byproduct of many industrial processes (eg petroleum refinery catalytic cracking and asphalt production) and because of the complicated atmospheric behaviour of formaldehyde. Ambient levels are also composed of direct or primary emissions as well as secondary emissions. The EPA ambient level estimate is based on the assumption that primary formaldehyde emissions are about 30% of ambient levels (of which 28% is due to motor vehicles) and secondary emissions account for 70% (of which 35% is due to motor vehicles).

It is important to assess formaldehyde levels from motor vehicles because of the expanded use of methanol-based additives in motor vehicle fuels. Methanol or methanol-based additives such as MTBE can lead to increases in both primary and secondary formaldehyde emissions.

6.3.2 Health Effects of Selected Mobile Source Air Toxics

According to some US studies, mobile source air toxics are responsible for the majority of estimated air pollution-induced cancer cases. The relative contribution of mobile source air toxics compared to stationary sources is largely dependent on the scope of the analysis. Estimating the potential incidence of health effects, such as cancer, that could result from exposure to air pollution is a complicated and uncertain process.

---

50 US Environmental Protection Agency, 1992, *Motor Vehicle-Related Air Toxics Study*, Public Review Draft, Washington DC.

51 R Derwent, 1992, *Submission to UK Expert Panel on Air Quality Standards*, Department of the Environment, London.

Estimates of the toxicity of the chemicals concerned, the amount that the source in question emits (in this case road transport), the circumstances of those emissions, and the potential exposure to populations in different situations must all be made. In addition, the quantification of health effects usually takes place on a chemical-by-chemical basis. The likely synergistic effects of chemicals on the human body are not as well known as single pollutants.

Despite these uncertainties, it is important to pay attention to this research and analysis because it is one of the major driving forces in environmental policies world-wide. Many countries around the world base their environmental legislation on just such analysis. The amount of pollution that is allowed to be emitted from a particular source is often determined using quantitative risk assessment. Risk analysis, whether based on 'maximally exposed individuals' or population 'incidence' estimates, is the method by which some policies and regulations are determined. The purpose of this section of the report is to bring to the attention of UK policy makers, industry and the public, the type of research that is taking place and, more importantly, the policy trends that are initiated by such research.

Figure 21: Relative Contribution by Source Categories to Total Estimated Cancer Cases per Year in the USA

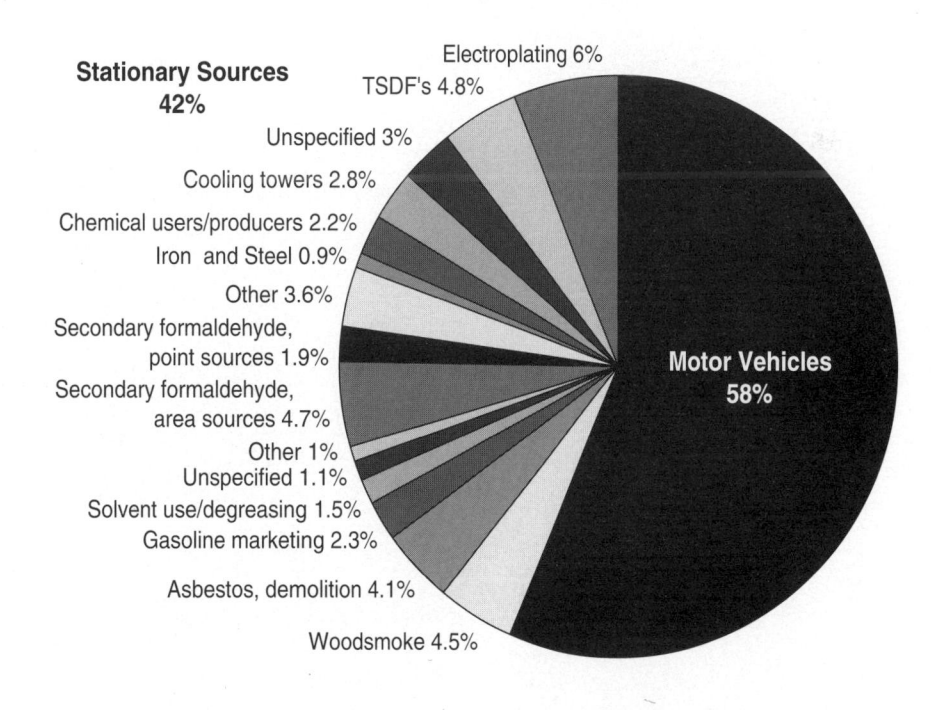

- Stationary Sources 42%
- Electroplating 6%
- TSDF's 4.8%
- Unspecified 3%
- Cooling towers 2.8%
- Chemical users/producers 2.2%
- Iron and Steel 0.9%
- Other 3.6%
- Secondary formaldehyde, point sources 1.9%
- Secondary formaldehyde, area sources 4.7%
- Other 1%
- Unspecified 1.1%
- Solvent use/degreasing 1.5%
- Gasoline marketing 2.3%
- Asbestos, demolition 4.1%
- Woodsmoke 4.5%
- Motor Vehicles 58%

Source: Control of Hazardous Air Pollutants in OECD Member Coubtries, OECD, 1991

A cursory review of some of the health effects of selected mobile source air toxics is given below. The review is not meant to be exhaustive. It is only a 'review of reviews', rather than an independent assessment of the validity of each body of substantial research underlying the summaries. It is designed to illustrate the necessity of considering air toxics in general, and mobile source air toxics in particular when formulating environmental public health policy. The majority of this information in the section unless otherwise cited is taken from EPA's recent draft review 'Motor Vehicle-Related Air Toxics Study'[52].

Benzene

Cancer Effects

Benzene is a known 'human carcinogen' as classified by both the International Agency for Cancer Research (Group 1) and the Carcinogen Assessment Group of the US EPA (Group A).

The EPA has calculated a cancer unit risk factor for benzene of 8.3×10^{-6} based on epidemiological studies. Updated studies provide additional and continued evidence of the carcinogenicity of benzene in humans. WHO epidemiological data gives estimates of excess leukemia deaths resulting from 30 years of occupational exposure to 1 ppm benzene ranging from 3 to 46 per 1000 exposed individuals. Lower exposure risk estimates range from 0.08 to 10 excess leukemia deaths per million exposed resulting from lifetime exposure of 1 µg/m^3.

Due to its carcinogenicity, there is no safe level for airborne benzene according to the WHO[53].

1,3-butadiene

Cancer Health Effects

EPA has classified 1,3-butadiene as a 'probable human carcinogen' based on tumors caused in animals and limited human data. The IARC has classified the chemical a 'possible human carcinogen.' A cancer unit risk factor of 2.8×10^{-4} has been set (after adjusting for an inhalation factor). EPA is reevaluating the cancer risk factor for 1,3-butadiene with recent animal studies indicating the occurrence of cancer in mice at lower concentrations than those used to

---

52 US Environmental Protection Agency, 1992, *Motor Vehicle-Related Air Toxics Study*, Public Review Draft, Technical Support Branch, Ann Arbor.
53 op cit note 43.

derive the present risk factor. Further work at lowering the uncertainty associated with risk factor estimation is necessary. It will involve isolating, with more precision, the fate of the chemical, the extrapolation of data from animal to human, and the specific tumor incidence data. The current risk factor should be considered as an upper bound estimate[54].

Non-cancer Health Effects

1,3-butadiene is also genotoxic and reproductive toxic. Exposure to very high levels is associated with eye, nose and throat irritations, respiratory paralysis and, at very high levels, death. Studies of rubber industry workers chronically exposed to the chemical suggest other potential health effects including heart disease, blood disease and lung disease. Other studies indicate that 1,3-butadiene appears to be a maternal toxin, fetotoxin, as well as a reproductive toxin.

Formaldehyde

"Formaldehyde is a colourless gas at normal temperatures with a pungent, irritating odour. It is the simplest member of the family of aldehydes and is the most prevalent aldehyde in vehicle exhaust and is formed from incomplete combustion of the fuel. Formaldehyde is emitted in the exhaust of both gasoline and diesel-fuelled vehicles. It is not a component of evaporative emissions"[55]. The use of a catalyst has been found to be effective for controlling formaldehyde emissions. Formaldehyde emissions are controlled to roughly the same extent as total hydrocarbon emissions with a catalyst[56].

Interest in motor vehicle formaldehyde emissions has increased as oxygenated fuels become more prominent. Oxygenates in fuel tend to increase formaldehyde emissions of catalyst-controlled and non-catalyst controlled vehicles.

Mobile sources are just one source of formaldehyde emissions. Other sources include the products that contain formaldehyde-based resins, such as plywood, particle board and counter tops. A study indicated that the high-exposure situations in the US were mobile homes (with their high percentage of particle board), offices (with their amount of furniture), and formaldehyde resin manufacturing occupational exposures. Mobile source exposure related inci-

---

54 op cit note 42.
55 US Environmental Protection Agency, 1991, *Locating and estimating air emissions from sources of formaldehyde*, Report no. EPA-450/4-91-012.c, OAQPS, Research Triangle Park.
56 US Environmental Protection Agency, Office of Mobile Sources, 1987, *Air Toxics emissions from motor vehicles*, EPA Report no. EPA-AA-TSS-PA-86-5, Ann Arbor.

dence was calculated to be between 2–6% of total health effect incidence of the chemical[57]. This study should be considered as very preliminary. It did not take into account potential increases in mobile source formaldehyde emissions due to expanded use of oxygenated fuels.

Cancer Effects

The EPA has classified formaldehyde as a Group B1 'probable human carcinogen' based on limited epidemiological and sufficient animal study evidence. Mutagenic activity evidence also exists. A recent EPA Office of Toxic Substances review of the formaldehyde risk assessment concluded that human studies released since the earlier 1987 study support the conclusions drawn earlier. The review did not alter the evaluation that 'limited' evidence does exist for an association between formaldehyde and human cancer, although the data do not conclusively demonstrate a causal relationship[58].

The IARC has classified formaldehyde as a group 2A carcinogen (i.e. 'probably' carcinogenic to humans). The California Air Resources Board also concludes that formaldehyde is a 'probable human carcinogen', and the US Occupational Safety and Health Administration ruled that formaldehyde should be considered a 'potential occupational carcinogen'.

Acetaldehyde

An aldehyde found in vehicle exhaust and formed through incomplete combustion, acetaldehyde mobile source emissions are estimated to be approximately 39% of total acetaldehyde emission inventories according to the EPA.

Cancer Effects

According to the EPA, sufficient evidence exists that acetaldehyde produces cytogenic damage in cultured mammalian cells. Based on this animal data, the chemical is classified as a 'probable human carcinogen'. The only epidemiological study involving acute exposure scenarios of acetaldehyde production workers showed a five times higher cancer rate among these workers than in the general population. Because of poor control, this study is considered inadequate to evaluate the carcinogenicity of acetaldehyde.

---

57 US Environmental Protection Agency, 1987, *Cancer Risk Due to Indoor and Outdoor Sources of Formaldehyde*, Memo from Charles Gray to Richard D Wilson.

58 US Environmental Protection Agency, 1991, *Formaldehyde Risk Assessment Update*, external review draft, Office of Toxic Substances, Washington DC.

Diesel Engine Exhaust

Cancer Effects

Based on animal experiments and two comprehensive human studies (a case-control study and a retrospective cohort study), the IARC concluded in 1989 that limited evidence for carcinogenicity existed and classified diesel engine exhaust as a 'probable human carcinogen'[59]. This view is shared by the UK Department of Health Committee on Carcinogens, and the US National Institute of Occupational Safety and Health.

In 1990, the EPA concluded that in regard to lung cancer, which is the endpoint used for evaluating human health risk, an excess risk was observed in three of seven human cohort studies and six of seven case-control studies. Based upon limited evidence for carcinogenicity of diesel engine emissions in humans, and supported by adequate evidence in animals, the EPA classified the compound as a probable human carcinogen.

PM-10

Based on several studies of the short-term and long-term effects (primarily non-cancer inhalation human studies) of PM-10, the EPA changed its ambient air quality standard in 1987 to concentrate on particulate matter of ten microns or less in size.

Since the PM-10 standard was set, a number of new epidemiological studies indicate that concentrations lower than the national ambient air quality standard of 150 $\mu g/m^3$ may be responsible for increased mortality[60]. An important fact in these studies was that daily mortality increased in a generally linear fashion according to increasing concentrations of PM. There also appeared to be no lower threshold and, in particular, effects were observed well below the current US health-based standard. In one study the mortality-PM association occurred even though ambient levels never bridged the national standard.

---

59 World Health Organization, 1989, *Diesel and Gasoline Engine Exhausts and some Nitroarenes*, Mono volume 46, International Agency for Research on Cancer, Lyon.

60 See for example:
J Schwartz, 1991, "Particulate air pollution and daily mortality in Detroit", *Environmental Research*, 56: 204–213;
J Schwartz and D W Dockery, 1992, "Increased mortality in Philadelphia associated with daily air pollution concentrations", *American Journal Epidemeology*, 135: 12–19; and
J Schwartz and D W Dockery, 1992, "Particulate air pollution and daily mortality in Steubenville", *American Review of Respiratory Disorders*, 145:600–604.

Concerning diesel versus petrol, some studies indicate that diesel cars can emit up to 100 times more particle per mile than unleaded petrol cars[61] and that when the mutagenic emission rates of petrol versus diesel vehicles are compared, the diesel vehicle emitted 45–800 times as much mutagenic activity per mile as the gasoline, catalyst-equipped vehicle[62].

Gasoline Particulate Matter

The potential health effects of gasoline particulate matter were also studied in the recent EPA draft assessment of the toxic health effects of mobile source air toxics. The small size of gasoline particulate matter makes it easily inhaleable into the deep part of the lungs.

Light-duty diesel engines emit some 30–100 times more particles than comparable gasoline engines equipped with catalytic controls; but due to the number of gasoline vehicles on the road, the ambient particulate loading may be a subject of concern.

If so, the health effects associated with gasoline particulates should be considered as similar to diesel particulate matter. At the time of the EPA draft review document, there were no official documents outlining or estimating the carcinogenicity of gas pm. The IARC has not developed a potency estimate for gas engine exhaust, but has listed it as a group B2 carcinogen (eg, 'possibly human carcinogenic').

Gas Vapours

Cancer Effects

The EPA estimated the carcinogenicity of unleaded gasoline[63] in 1987 (a summary of research conducted up until 1985). Based on the weight of evidence at the time, the EPA classified gasoline vapour as a group B2 probable human carcinogen. Animal and human epidemiological work since 1987 calls the original EPA finding into question. Reviews of epidemiological studies did not provide enough support for an association between exposure to petrol hydrocarbons (excluding benzene) and renal cancer[64]. The draft mobile air

---

61 J Lewtas, 1991, *Carcinogenic risks of polycyclic organic matter (POM) from selected emission sources*, Health Effects Research Laboratory, Research Triangle Park, N C.

62 L C Claxton & M Kohan, 1987, "Bacterial mutagenesis and the evaluation of mobile-source emissions", in *Application of short-term bioassays in the fractionation and analysis of complex environmental mixtures*, pp 299–317, M D Waters, S S Sanduh, J L Huisingha, L C Claxton, S Nexnow, editors, New York.

63 US Environmental Protection Agency, 1987, *Evaluation of the Carcinogenicity of Unleaded Gasoline*, EPA Report No. 600/6-87/001.

64 J M Harrington, 1987, "Health experience of workers in the petroleum manufacturing and distribution industry: a review of the literature", *American Journal of Industrial Medicine*, 12:475–497.

toxics review study did not attempt to extrapolate unit risk and exposure estimates derived from earlier EPA work to estimate gasoline vapour cancer incidence.

Non-Cancer Effects

Low and/or acute gas vapour exposure can cause respiratory tract irritation, nausea and headaches. Higher concentrations may cause pulmonary edema, heart damage, central nervous system depression and other nervous system disorders[65]. Chronic inhalation may cause dizziness, nervousness, limb pain, weight loss, and fatigue, among other disorders. Severe exposure can result in irreversible effects such as aplasticanemia and leukemia[66].

6.4 Human Health Risk Estimates Due to Motor Vehicle Toxic Air Pollutant Emissions

There have been several studies estimating the potential increased incidence in cancer cases due to exposure to motor vehicle air pollution emissions. The most comprehensive analysis has been completed recently by the US EPA. It incorporated up to date emission factors, exposure scenarios, and projected changes in emissions and exposures due to the implementation of the various new emission control programs of the US Clean Air Act Amendments of 1990[67].

Table 21 below summarizes the results of this project which estimated increased cancer incidence for benzene, formaldehyde, 1,3-butadiene, acetaldehyde, diesel particulate matter, and gasoline particulate matter under various emission control scenarios.

- Scenario 1 estimated the incidence that would result due to the implementation of the new motor-vehicle requirements of the Clean Air Act.
- Scenario 2 estimated the incidence involving the expanded use of reformulated gasoline, in addition to the controls of scenario 1.
- Scenario 3 expanded the adoption of California emission standards to other parts of the US (the Northeast and the worst ozone polluted areas).

The study estimated cancer incidence for the years 1990, 1995, 2000, and 2010. It should be noted that the quantification of potential cancer incidence is

---

65 Material Safety Data Sheets, Occupational Health Services Inc, 1985; Howell Hydrocarbons Inc, 1987; Sun Oil Company, 1989.

66 US Environmental Protection Agency, 1980, *Facts and issues associated with the need for a hydrocarbon criteria document*, internal document, Office of Research and Development, Environmental Criteria and Assessment Office.

67 op cit note 50.

Table 21 US Mobile Source Cancer Study: Summary of Emission Factors, Exposure Estimates, and Nationwide Cancer Incidences[a]

| | | \multicolumn{3}{c|}{Benzene} | \multicolumn{3}{c|}{Formaldehyde} | \multicolumn{3}{c|}{1,3-Butadiene} |
		EF	Exp[b]	CC	EF	Exp[c]	CC	EF	Exp	CC
1990	Base Control	0.008	2.36	70	0.041	0.56	26	0.016	0.42	418
1995	Base Control	0.047	1.40	43	0.023	0.37	18	0.009	0.28	286
	E R Gas Use	0.041	1.20	37	0.025	0.39	19	0.009	0.27	283
2000	Base Control	0.035	1.10	35	0.016	0.26	13	0.007	0.23	242
	E R Gas Use	0.030	0.98	31	0.017	0.26	13	0.007	0.22	235
	E A Cal Standards	0.030	0.98	31	0.017	0.26	13	0.007	0.22	236
2010	Base Control	0.028	1.05	35	0.014	0.25	13	0.007	0.25	280
	E R Gas Use	0.025	0.93	31	0.014	0.27	14	0.006	0.24	267
	E A Cal Standards	0.023	0.84	28	0.014	0.25	13	0.006	0.23	256

based on a modelling exercise. Its relevance for the UK is in highlighting pollutants of concern for further research or control and in evaluating road transport pollution control policies. This recent study is particularly important because it evaluates changes in vehicle emissions due to the adoption of new types of vehicle fuels, specifically 'reformulated' petrol. Reformulated petrol is discussed in detail in Section 8.3.

Study Results

- The base-line mobile source cancer risk in 1990 was estimated at 720 cases. This drops to 470 in 1995, 370 in 2000 and increases to 391 in 2010 (due to increases in vehicle miles travelled which are no longer being offset by decreases in vehicle emission factors).
- For the years studied, 1,3-butadiene was responsible for the majority of the cancer incidence – from 58 to 72% of the total. This is due to the chemical's relatively high unit risk factor.
- Gasoline and diesel particulate matter contribute roughly the same to aggregate increases in estimated cases of cancer – the combined risk is from 16 to 28% of the total.
- Benzene is responsible for roughly 10% of the total estimated incidence and aldehydes (formaldehyde and acetaldehyde) for approximately 4%.

Table 21 continued

		Pollutant								
		Acetaldehyde			Diesel Particulate Matter			Gasoline Particulate Matter			Total
		EF	Exp[d]	CC	EF	Exp	CC	EF	Exp	CC	CC
1990	Base Control	0.012	0.51	4.0	0.067	1.80	109	0.020	0.51	93	720
1995	Base Control	0.007	0.34	2.8	0.036	1.05	66	0.011	0.29	54	470
	E R Gas Use	0.007	0.34	2.8	–	–	66	0.011	0.28	53	461
2000	Base Control	0.005	0.25	2.1	0.019	0.60	39	0.008	0.20	39	370
	E R Gas Use	0.005	0.25	2.1	–	–	39	0.007	0.19	38	358
	E A Cal Standards	0.005	0.25	2.1	–	–	39	0.007	0.19	38	359
2010	Base Control	0.004	0.25	2.2	0.010	0.39	27	0.006	0.17	34	391
	E R Gas Use	0.004	0.25	2.2	–	–	27	0.006	0.16	32	373
	E A Cal Standards	0.004	0.24	2.1	–	–	27	0.006	0.15	31	357

Notes:
- EF   Emission Factor (grams/mile).
- Exp   Exposure ($\mu g/m^3$).
- CC   Estimated Cancer Cases.
- E R Gas Use   Expanded Reform Gas Use.
- E A Cal Standards   Expanded Adopted California Standards.

a Cancer incidence estimates are based on upper bound estimates of unit risk, except for benzene. Unit risk estimate for benzene is based on human data. Cancer incidence estimates are meant to be used in a relative sense to compare risks among pollutants and scenarios, and to assess trends.
b Exposures given are nationwide estimates.
c Formaldehyde exposures have been adjusted downward by 50% to agree with ambient data.
d Nationwide acetaldehyde exposures have been adjusted upward by 40% to agree with ambient data.

Source: Motor Vehicle-Related Air Toxics Study, EPA, 1992

- The expanded use of 'reformulated' gasoline (which has by definition 16.5% less VOC and air toxic emissions) resulted in between 2 to 9% reduction in incidence depending on the scope of implementation of the programme.
- Oxygenated fuels provided some incidence reduction in terms of reduced emissions of aromatic compounds such as benzene, and a slightly increased risk due to increases in aldehyde emissions.

Summary Points

- Mobile source air pollution contributes to estimated cancer incidence in the US. This is so even in a base-case scenario that acknowledges that a major proportion of the US fleet is equipped with either catalysts or 3-way catalysts, that unleaded petrol in 1990 represented over 85% of the market and that the average speed (and hence emission factors) of the US fleet is lower than that in the UK.
- Mobile-source related cancer incidence can be reduced by implementing both exhaust controls and improving the quality of motor vehicle fuels. A 35% reduction in estimated cancer cases was achieved by both reducing the combined HC + NO_x emission standard for 80% of the US new car fleet by approximately 50% and by expanding the market share for Federal Phase I reformulated gasoline from 0 to 18% of the total US market.
- It is not possible to ascertain from the study results available in the draft report what percentage of the decline in cancer incidence was attributable to tailpipe controls or to reformulated gasoline usage. Expanding the market share of reformulated gasoline from 18 to 50% from 1990 to 1995 while holding constant the tailpipe controls seems to result in an additional decline in cancer incidence of only 1%.
- The lack of further cancer incidence reduction with the expansion of the reformulated gasoline market share bears further examination. Reform, with an average 16.5% fewer VOC and air toxic (benzene, 1,3-butadiene and formaldehyde) emissions levels and a benzene content limit of 1% by volume, should result in lower toxic emissions and lower incidence. In addition, the use of reformulated fuels across the whole fleet should show proportionally greater benefits than technical controls applied to new cars only.

Implications and Future Directions for the UK

In order to extrapolate the results of the US study to the UK, certain assumptions and adjustments would have to be made. If one assumed that the unit risk factors derived in the US study were applicable to the UK, then UK exposure estimates and UK emission factors for mobile source vehicles that are relatively less controlled than the US would have to be developed. UK emission factors would have to account for both the difference in fuel quality between the US and the UK and vehicle emission control technology. Vehicle use and fleet composition would also have to be accounted for. Without conducting any of this analysis, some observations can be made from a relative point of view.

- The vehicle emission factors for a UK 1990 base case are much higher than for the US 1990 base case. The UK fleet is operating primarily without the benefit of catalyst technology, with a higher proportion of leaded fuel, a higher sulphur content fuel, a higher volatility pressure fuel and a higher average benzene content.
- The UK fleet also uses proportionally more diesel fuel than the US, so the cancer incidence attributable to diesel particulate matter would play a proportionally greater role. The UK fleet also drives at a faster average speed, increasing individual emission factors.
- In conclusion, the per-vehicle cancer incidence estimated in the US study for 1990 can only be greater for the 1990 UK fleet, assuming equivalent unit risk factors and similar population exposure estimations. Otherwise, all the lessons from the US experience should apply to the UK as well. These include:
- Mobile source air emissions increase the incidence of cancer in the general population and the incidence of non-cancer health effects. Compared to other sources of air pollution, mobile sources are estimated to contribute to at least 50% of total air pollution-induced cancers.
- Mobile sources contribute the greatest amount to the ambient concentration of known carcinogens – specifically, benzene. Benzene ambient concentrations can be reduced by limiting the amount of benzene in fuel and by the application of catalytic converters to the fleet.
- In the UK, urban populations are at greater risk than rural populations from mobile source air toxic emissions.

Quantitative risk analysis, if applied to transport health effects as well as to other environmental problems, can be of a great assistance in evaluating research programmes and setting policy priorities.

It should be pointed out that an analysis like the one reviewed above, is unlikely to take place in the near future in the UK. Numerous toxicologists in the UK do not favour the sort of quantitative risk assessment that is the basis of the US studies and numerous US environmental regulations[68]. While such a position may act to preserve a certain scientific integrity on the part of the British toxicologist community, it does not assist the UK government and environmental policy makers in evaluating policy trends in the rest of the world that are

---

[68] UK Committee on Carcinogens, 1991, *Guidelines for the Evaluation of Chemicals for Carcinogenicity*, HMSO, London.

based on such quantitative analysis. In addition to this official view on quantitative risk assessment, there is still doubt on the part of UK policy makers concerning the significance of road transport air pollution health problems[69].

6.5 Other Mobile Source Air Toxic Information and Studies

6.5.1 VOCs and Air Toxics

Relatively innocuous VOCs, when photo-oxidized by sunlight in the presence of nitrogen oxides (NO_x), have been observed to produce air toxics. For example propylene and toluene, two non-mutagenic volatile organics have been shown to produce mutagenic products when oxidized in the presence of NO_x. The photo-oxidation products that produce this bioactivity have been tentatively identified as oxygenated and nitrated products. Similarly, diluted mixtures of irradiated automobile emissions were found to produce significantly increased mutagenic activity compared to non-irradiated mixtures.

Thus, VOC reduction strategies can be beneficial for control of secondarily formed toxics[70].

Implications for the UK

- The UK should assess the implications on ambient levels of air toxics from the VOC control strategy that is currently under preparation and determine what actions would assist in optimizing control of the more hazardous species of VOCs.

6.5.2 EPA's Integrated Air Cancer Project

A major research programme recently undertaken in the US is attempting to measure and allocate mutagenicity of ambient air samples to various sources[71]. The study focuses on the products of incomplete combustion, including polycyclic organic matter (POM). Preliminary study results from Raleigh, North Carolina, Albuquerque, New Mexico and Boise, Idaho indicated the following:

- Mobile sources contributed between 20–56% of the extractable organic material (EOM) sampled[72].

---

69 For example, the DoT's Chief Medial Advisor believes that "there is no certain association between vehicle emissions and human cancer." J F Taylor, 1993, Comments on Foresight Draft Report, Department of Transport, London.

70 K D Mason et al, 1991, *Clean Air Act Amendments of 1990: Benefits Assessment, Draft Policy Paper*, US Environmental Protection Agency.

71 L Cupitt and T Fitz-Simons, 1988, *The Integrated Air Cancer Project: Overview and Boise survey results*, APCA Publication VIP-10; EPA Report No. 600/9-88-015.

72 Extractable organic material (EOM) is a measurement of particulate organic material and typically accounts for about 60% of the fine particles collected in sampling at residential and mobile sites.

- These results support earlier conclusions from the EPA's 'Six-Month' cancer air toxics study which attributed approximately 50% of the incidence to mobile sources[73].
- The mobile source contribution would be greater in cities larger than those studied.
- The mutagenic potency of matter from mobile sources was roughly three times higher than from woodsmoke, and the lifetime unit risk was 2.5 times greater.

6.5.3 Other National Air Toxics Control Programs and Research

Based on the work of the Air Management Policy Group of OECD's Environment Directorate, reviews are being conducted of air toxic control programmes around the world. Although most of the review material presented in draft OECD documents does not discuss specific mobile source air toxics, an overview of the programmes is useful to gain an appreciation of the work being conducted in other countries and the air quality management goals and targets that are being set as a result. When completed, the OECD project will be a good overview of the type of information, policies and programmes that are being considered by other countries. Information presented below is taken from work in progress[74].

France

France currently has no specific programme for hazardous air toxic control and little inventory of air toxics. It is considering developing a more active programme along the lines of TA Luft in Germany. Identified priorities to date (as part of the National Plan for the Environment issued in 1990) include: reduction of halogenated organic compounds by 1995 (no % reduction stated); and the assessment of new material usage, such as polymers, ceramics, carbon fibres, memory alloys, selenium, beryllium and the environmental byproducts of new fuels.

Germany

Through TA Luft, Germany expects VOC reductions of around 40% from 1986 to 1995, with equal amounts of heavy metal reduction. Evaporative emission controls of toxics and halogenated organic compounds are cited for

---

73 US Environmental Protection Agency, Office of Air Quality Planning and Standards, 1990, *Cancer Risk from Outdoor Exposure to Air Toxics*, EPA 450/1-90-004a, Research Triangle Park, North Carolina.

74 OECD Environment Directorate, Air Management Policy Group, 1992, *Control of Hazardous Air Pollutants in OECD Member Countries: Summary Report of Seven Country Surveys*, Draft, Paris.

improvement[75] and proposals are being developed to set emission standards for a targeted group of carcinogens, where the ultimate risk level goal being considered is 4×10^{-4}. Measures taken for other purposes have had beneficial air toxics results: tetrachloroethylene and perchloroethylene have been reduced from dry cleaning controls and petrol evaporative emissions have been regulated through controls on storage and handling.

Japan

A new programme is being implemented by the Environmental Agency, whereby the risk of substances will be assessed and necessary control measures designed. To date, attention has centred on cadmium, lead, chlorine, fluorine, hydrogen fluoride and silicon fluoride.

The Netherlands

Risk limits have been established for 20 priority substances. The Netherlands is the only country that has established ambient air quality standards for a series of air toxics. In addition, The Netherlands pursues a source-specific emission standard method. Future policy will be source and sector-oriented and will use emission guidelines developed for this purpose. The emission standards are technology-based rather than risk-based. In this respect, the current Netherlands programme resembles the new US air toxics programme. The new US air toxics programme will establish technology-based standards for a prescribed number of pollutants and sources of those pollutants. After a level of technological control has been in place, an evaluation of the "residual" risk that remains is made and additional control is required if necessary.

Sweden

Overall cancer incidence in Sweden from air pollution is estimated to be between 300–2000 cases per year. The substances considered were VOCs like benzene, as well as particulate matter and PAH. The urban air toxic annual population risk of 1×10^{-4} was cited. In addition to attaining the air quality standards for conventional ambient pollutants, Sweden seeks a substantial reduction in the total estimated cancer risk resulting from air pollution. A 90% carcinogen reduction goal has been established and motor vehicles in urban areas are being targeted first. Sweden is taking the lead in an international effort to reduce or ban certain persistent organic compounds by the year 2000, in conjunction with other eight other European countries. Goals of reduction

---

75 Being considered are arsenic, asbestos, benzene, cadmium, benzo (a) pyrene, diesel particles, and 2,3,7,9-TCDD.

include 15 pollutants by at least 50% by 1999, and emissions of mercury, cadmium, lead and dioxins by 70% by 1999 from a 1985 base-line.

Implications and Future Directions for the UK

- A range of air toxic air pollution control programmes exist for the UK to analyze and assess. As these programmes proliferate, information about toxic emissions from various sources, including mobile sources, will increase and an air toxics control strategy could be implemented in conjunction with existing programmes.

6.6 Air Toxic Exposure Situations

6.6.1 In-Vehicle Passenger Air Pollutant Exposure

In recent years, the level of exposure to vehicle occupants and to certain labourers, (such as service-station workers, parking garage attendants and toll-booth operators) have been the subject of study. In the case of in-vehicle emission levels, studies indicate greatly elevated air pollution levels relative to ambient conditions. A comparative look at several of these studies by the UK Urban Air Quality Review Group concluded that in-vehicle CO and lead concentrations were on average about five times higher than urban background locations[76].

A detailed look at one of these studies which analyzed in-vehicle VOC emission levels under different driving conditions in comparison to vehicle exterior concentrations, fixed-site measurements and sidewalk measurements reveals the following[77]:

- VOC and CO concentrations were higher in vehicles than at fixed sites (100–300 feet away from roadways).
- Fixed-site ozone levels were higher than in-vehicle levels which were only 20% of the ambient ozone concentrations.
- On average, the in-vehicle VOC concentrations were approximately six times higher than the VOC levels at fixed sites and CO was approximately 4.5 times higher than ambient CO measurements.
- The median in-vehicle concentration for 1,3 butadiene was 2.9 $\mu g/m^3$ and 10.7 $\mu g/m^3$ for benzene.
- The toluene concentration in vehicles was approximately 12 times higher than fixed-site ambient levels.

---

[76] op cit note 9.

[77] C Chang-Chuan, H Ozkaynak, J D Spengler, & L Sheldon, 1991, "Driver Exposure to Volatile Organic Compounds, CO, Ozone, and NO_2 under Different Driving Conditions", *Environmental Science and Technology*, 25:964–972.

- A significant amount of vehicle-related VOC exposure is associated with roadway type and traffic conditions, according to statistical analysis of the data. Urban driving conditions contributed to the highest in-vehicle conditions, interstate (motorway) conditions the next highest, and rural conditions the least. The ratio was 10/6/1 (urban, interstate, rural).
- In-vehicle VOC levels were lowest when mobile air conditioning was used.

6.6.2 Street Canyon and Kerbside Concentrations of Pollutants

In addition to elevated concentrations of air pollutants within vehicles, kerbside monitors indicate elevated concentrations of pollutants when compared to urban ambient background levels.

One study of two sites in Central London indicated that nitrogen dioxide concentrations ranged from 80–90 ppb in the centre of the road, 50–60 ppb at the kerb, and 40–50 approximately three metres from the kerb[78]. This compared to a local background concentration of 30–40 ppb.

Further analysis of this information indicated that:

- Annual mean concentrations were on average 5–10 ppb higher than local background levels at 10 metres from a busy urban roadway kerb;
- 7–15 ppb higher at five metres from the kerb; and
- 15–30 ppb higher at around one metre from the kerb.

Implications and Future Directions for the UK

- As kerb-side and in-vehicle pollutant level monitoring increases and risk estimations can be made, urban mobile source pollution reduction measures (traffic planning, fuel substitutions, etc) will be able to be evaluated for their contribution to improved urban public health.

---

[78] The Second Report of the United Kingdom Photochemical Oxidants Review Group, 1990, *Oxides of Nitrogen in the United Kingdom*, prepared at the request of the DoE, London.

7 The Regulation of Motor Vehicle Air Pollution

In order to ascertain the future direction of motor vehicle emission control policy, it is useful to identify the major current trends in that development. The following discussion primarily concerns passenger car policy development in the US. The relevance of these developments for the EC and the UK, and the direction in which they may lead, will be discussed later.

7.1 A Short History

In the 1950s, Los Angeles's smog was conclusively linked with motor vehicle emissions and California imposed crankcase emission standards in 1960 which were followed soon after by tailpipe exhaust emission standards. The US Congress enacted a research and technical assistance program for motor vehicle-related pollution with the Motor Vehicle Act of 1960. As evidence accumulated linking air pollution to heart and chronic respiratory diseases, the US federal role was expanded in 1963. The 1963 law required the Federal Dept of Health, Education and Welfare to gather the scientific evidence to set criteria for air pollution standards based on health effects. This evidence, and a growing federal role, resulted in the Motor Vehicle Air Pollution Control Act of 1965, where the government was directed to set emission standards based on "technological feasibility and economic costs." The standards that resulted in 1967 were soon updated by the US's first truly technology-forcing pollution control standards in 1970.

Under the 1970 Clean Air Act, Congress required auto manufacturers to reduce HC and CO pollution by 90% within five years and NO_x within six years. The US auto industry reacted to the technical challenge by colluding to prevent the introduction of emission control strategies. Their action was eventually investigated by the US Justice Department, a case which ended with General Motors, Ford and Chrysler signing a consent decree to stop this collusion.

The court case, combined with competition from abroad, other new government regulations and market forces, acted together to move the industry toward embracing regulation of motor vehicle pollution. Eventually, an incremental technological innovation path resulted that led to the development and use of the catalytic converter rather than basic engine redesign.

This reaction to the technical challenge posed by emission control was due to a number of factors. Normally, the industry's innovative efforts were channelled into enhancing performance. Emissions reductions ran counter to performance enhancement. The industry at that time was not able to identify any clear consumer demand for either 'cleaner' vehicles or more fuel efficient vehicles and the adverse public health effects of air pollution were not readily appreciated by the industry. Emission control strategies were thus consigned to 'after-design' and the main objective of engineers working to control emissions was one of cost-minimization. The US industrial response differed from that of the UK auto manufacturers, who invested their efforts in lean-burn engine design. These efforts were thwarted by a lack of sufficient technical progress. They were eventually overcome by catalyst-based emission control policy trends which started in the US.

7.2 Notable Trends In US and World Passenger Car Air Pollution Control

7.2.1 The Environmental Goal of Motor Vehicle Air Pollution Control

Motor vehicle emission control policies have initially concentrated on so-called conventional air pollutants – the precursors of tropospheric ozone or smog (HC and NO_x) and other commonly known pollutants such as lead, carbon monoxide, and particulate matter. The initial concentration was on hydrocarbon emissions because of their contribution to the formation of smog. Later, regulatory emphasis on NO_x emissions developed because of the increased understanding of the relationship between NO_x emissions and smog. In addition, the harmful effects of NO_2 itself became a factor for increased NO_x control. The contribution of motor vehicles to other regional air pollution problems, such as acid rain, has also been acknowledged through regulation.

Recently, global pollution concerns of stratospheric ozone depletion and global warming have expanded the scope of concern surrounding road transport emissions. CFCs and HCFCs used in manufacturing and operation have become regulated and the heightened desire for reduced CO_2 emissions from transport has joined earlier oil-price shocks and domestic energy security concerns to reinforce the need for increased energy efficiency in the transport sector.

7.2.2 The Changing Scope of Motor Vehicle Pollution Problems

The above trend in environmental goals is matched by a changing scope of concern. Initially it was easy to say that motor vehicle pollution was considered a local problem, nationally distributed. As the variable severity of air pollution problems across the country became known and as air chemistry models developed, mobile source air pollution problems (smog in particular) took on a regional definition. Long-range ozone precursor transport and ozone formation became a concern. Global air pollution problems, attributed in part to mobile sources, have been the most recent scope of concern.

The next likely scope of concern will again be local conditions, but on a 'micro' scale. As levels of air pollution, particularly hazardous air toxics, become assessed from a 'micro' point of view (eg within-vehicle measurements, kerbside, urban–rural), policies may develop to address these micro-environment high pollution areas. To date, the elevated pollution levels and health effects of certain micro-environmental settings have not been the subject of much policy attention. Developments in indoor-air pollution research and in urban meteorology should stimulate a demand for such a policy focus, given the increased information.

It is likely that the response to these micro-environmental conditions will initially be met with planning and occupational standards rather than with technological standards aimed at the entire fleet.

7.2.3 From Ambient Air Quality Standards to Vehicle Emission Standards

Concern for human health is the driving force behind the current technology-based mobile source emission standards. The origin of tailpipe emission standards in California and the US as a whole were designed to help alleviate local and regional ambient air pollution problems (primarily smog) and to eventually help attain ambient air quality standards. These ambient standards for particulate matter, sulphur oxides, carbon monoxide, nitrogen dioxide, ozone and lead were initially determined through a review of the health effects research of these pollutants. Ambient levels and averaging periods varying from one hour to one year were established. The levels set in the United States were not 'no effects' levels or thresholds, but levels designed to "protect human health with an adequate margin of safety".

However, ambient air quality standards alone do not determine emission standards for various contributing sources. Initial and subsequent motor vehicle standards were implemented to reduce the contribution of motor vehicle pollution to these ambient problems. The exact amount of pollution reduction required by these new technology standards was not determined according to the environmental effect of the standards, but rather by some judgement of technical feasibility. The judgement of technical feasibility was itself based on a combination of technical, economic and political assessment and manoeuvering.

The evolution of vehicle-based emission standards has also been a result of many other factors. Certainly, as technical capacity improved, regulations were amended to incorporate the new advances. In addition, the intractability of the pollution problems themselves created a constant pressure for additional motor vehicle reductions. An initial scepticism of the public and the government towards automakers' claims of technical infeasibility permeated any subsequent discussions about improvements in standards.

7.2.4 From National to Local Air Quality Management Strategies

Another important development in the history of vehicle emission control policy was the development of local as opposed to national air quality management areas. Because of a lack of progress in meeting the national standards through national measures alone, *local air quality areas* were established, *local air quality improvement plans* were developed, and local measures (in addition to national measures) were implemented.

This change in air pollution management had several important effects.

- It created a demand for the development of local and regional air quality models. These in turn created a demand for improved information regarding what sources were contributing to emission reductions and eventually estimations of the environmental effects of proposed emission reduction policies. Thus, eventually, the specific contribution of motor vehicles to air quality was assessed and quantified.
- Local air quality management areas implemented sanctions and penalties for those local areas that did not meet national goals, which kept up the pressure for further national solutions (manifested in tighter tailpipe standards) and for creative local solutions (eg, looking past the mobile source to other emission sources).
- The local air quality goals and areas also tended to localize and increase pressure for action. Instead of "the US having 256 exceedences of the national standard for ozone in 1983", for example, the air quality compliance took on local significance such as, "San Francisco violated the national air quality standard 19 times in 1983", and so on.

Thus, local responsibility for attaining nationally set standards is a major driving force in regulatory development, at least in the US.

7.2.5 From Tailpipe to Fuel Distribution: The Changing Focus of Vehicle Regulation

In addition to an evolution of tailpipe emission control stringency, other sources of vehicle pollution have come under regulation over time. Initially the tailpipe itself was the sole target for vehicle regulations. Next, policies were put in place to make sure that the emission control technology was operating correctly – both as designed and as used. Vehicle *inspection and maintenance programs* were implemented to maintain the performance of the emission control system during use. In addition, vehicle manufacturers had to test and certify their vehicle's emission control performance and *guarantee* a level of performance for a defined period (usually expressed in terms of number of miles driven). These emission performance criteria were backed up by the force of *vehicle recall legislation* in the US. The most recent development in terms of insuring the performance of vehicle emission control systems is the requirement for on-board diagnostic computers to note and record the periods and nature of emission and engine performance, store this information electronically, and incorporate it into vehicle maintenance systems.

After the tailpipe controls had been in place, there was still a need to improve overall emissions performance. Other sources of vehicle emissions came to be better understood and eventually regulated. *Evaporative emissions* created during the normal course of vehicle warming and cooling and *refuelling emissions* (emissions purged from the petrol tank of the vehicle during refuelling and emissions escaping from the nozzle itself) were both identified as substantial vehicle emission sources and eventually controlled.

7.2.6 The Fuel Side of the Equation

Another major development in the evolution of vehicle pollution control came with the control of lead content in gasoline. The excellent progress in reducing ambient lead levels as a result limiting lead fuel content, brought to the attention of policy makers the potential benefits of addressing the fuels side of the motor vehicle pollution equation.

This attention has led in two main directions. Initially the quality of the fuels (diesel and petrol) was assessed and subject to increased specificity. Fuel additives, sulphur content, reactivity, and a whole host of primarily environmental criteria have been added to the existing fuel performance criteria. The environmental performance of vehicle fuels is now being evaluated with a number of environmental goals in mind (conventional pollution reduction, global pollution reduction, and local toxic pollution reduction). The US reformulated gasoline program and other fuel quality developments are reviewed in detail in Chapter 8 below.

Attention has also focussed on developing alternatives to diesel and petrol themselves. Alternatives have been explored in many countries and required in a few. The large scale Brazilian ethanol programme is one example, and investments in diesel substitutes for urban transportation, such as compressed natural gas, is another. These two trends are not wholly separate. Some fuel substitutes, such as oxygenated fuels (ethanol and methanol), are being investigated as both an environmental-performance enhancing additive and as a complete substitute to petrol.

7.2.7 From Passenger Cars to Lorries to Fleets

Passenger cars have been the initial source of motor vehicle pollution policy, followed by attention to lorries and buses. This is primarily due to the proportionally greater contribution that passenger cars have made to atmospheric pollution than less numerous HGVs. As the passenger cars contribution to total road transport reduction declines, relative to HGVs, attention will continually shift toward lorries. For example, the contribution of HGVs to UK total road transport NO_x emissions will increase from 34% in 1990 to 48% in 2010.

The increased attention on HGVs, combined with developments in fuel quality and substitution regulation, culminate in regulations requiring the use of alternative fuels in urban lorry and bus fleets. The centrally-fuelled nature of urban fleets lends itself to the introduction of alternative fuel vehicle programs.

7.2.8 From Vehicle Performance to Vehicle Use

Similar to the attention devoted to ensuring that the emission control systems of vehicles are operating according to specifications (inspection & maintenance programs, certification & testing, recall), the manner in which vehicles are driven is also becoming recognized as a contributor to emissions performance. While not regulated on environmental grounds alone, vehicle speed is being regulated for other reasons (safety and fuel economy) and the environmental benefits of certain driving conditions are being quantified and assessed.

Where regulation of vehicle driving coincides with other policy objectives (such as uniform freight handling and safety objectives for EC freight), the environmental and energy benefits of regulated lower speeds are increasingly being looked at by policy makers. Likewise, the impact of congestion on driving conditions and environmental performance are increasingly being analyzed and addressed through policy.

7.2.9 Vehicle Emission Regulation to Transport Planning

The most obvious and necessary trend in motor vehicle pollution control policy is towards managing the demand for transport and planning its use with environmental quality in mind. From restrictions on vehicle use of various sorts (eg high-occupancy lanes on motorways, lorry bans, parking restrictions) to expansion and improvements in substitute modes of transport (eg expansion of pedestrian zones in urban areas, dedicated bus and bike lanes, improved public transport), the ultimate solution to transport-related environmental problems lies with managing and planning transport demand and use. Transport planning efforts to reduce pollution have not had the attention and success to date as have technical control measures. Vehicle emission performance improvements should therefore not be abandoned. Successful incorporation of ever increasing transport demands into an environmentally sustainable future will require progress along both fronts. This report addresses the vehicle performance side of the equation and does not attempt to evaluate the contribution of transport planning to improved environmental quality.

7.2.10 From 'Command and Control' Regulations to Market Forces

Traditionally, road transport air pollution has been addressed through specific regulations on vehicle emissions and fuel quality. New environmental issues associated with motor vehicles, such as CO_2 emissions and global warming, lend themselves to more market-based policy methods. Increased fuel taxes and road pricing to reduce demand for passenger car transport are being implemented or experimented with in many countries, including the UK. Emissions trading policies, similar to those enacted to address acid rain emissions in the US, are being contemplated for mobile sources.

7.2.11 From Mobile to Stationary Sources

To date, automobiles and road transport in general have been identified as the main sources of some major air pollution problems (eg, lead, ozone). As these sources become controlled, as further incremental improvements become more difficult and expensive, and as some of these environmental problems persist due to the growth in vehicle miles travelled, there is a trend towards identifying stationary contributors to air pollution problems traditionally associated with road transport.

In Southern California, for example, there is serious consideration being given to mobile-stationary source 'emission trading'. Relatively expensive incremental reductions in vehicle emissions would be substituted with relatively less expensive reductions in stationary source pollution. Another example concerns NO_x pollution. As road transport sources of NO_x are controlled, off-road sources (eg agricultural tractors) make up a larger part of the NO_x inventory and became controlled sources. Eventually, non-road transport sources of NO_x, such as airplanes, and stationary sources, such as internal combustion engines and boilers, will be controlled. There is a noticeable trend toward considering stationary and mobile source pollution control measures together in an attempt to optimise control economics.

Another example of this convergence of stationary and mobile source control policies is illustrated by the control of gasoline vapours. Here, controls have developed to cover the entire production, distribution and use phases of gasoline. Bulk gasoline storage terminals at refineries, gasoline barge and tank loading operations, service station tank loading operations, and vehicle refuelling itself have all come under regulation. It is interesting to note that in the US, the purpose of initial refinery vapour controls was for benzene control due to its carcinogenic properties. The purpose of vehicle refuelling and gasoline handling and distribution controls was primarily for the control of hydrocarbons as precursors of tropospheric ozone. Not only have the stationary and mobile source control policies merged, but different environmental quality goals have also merged with interesting secondary effects.

7.3 Global Warming: The Merging of Fuel Economy and Environmental Policy

The emergence of global warming as a potential environmental danger has created one of the greatest challenges for the road transport sector of the economy. Reducing CO_2 and other greenhouse gas emissions from road transport at the same time as reducing "conventional" pollutants will require advancements along several fronts. Certainly there will be a renewed demand by governments for improvements in fuel efficiency. Global warming will join domestic energy security and other reasons to create continued pressure to improve vehicle energy efficiency.

In the US, the pressure to reduce transport's CO_2 contribution will most likely result in incremental increases in the fuel economy standards for vehicles and for the expansion of such fuel efficiency standards to other classes of vehicles not currently covered by this legislation. Yet even this development will not be easy. Relatively low fuel prices in the US, combined with the cavalier attitude of former Administrations towards global warming, and an overall lack of formal connections between the energy and environmental spheres of policy, all act to reduce the likelihood that corporate average fuel economy standards (CAFE) will be tightened on the basis of global warming arguments alone. And as recent experience has shown, it is highly unlikely that petrol demand will be limited through price increases brought about by substantially increased fuel taxes.

In Europe, it is much more likely that road transport's CO_2 contribution will be addressed from two perspectives. Increasingly, fuel taxes are being used by governments to try to curb demand and foster technological efficiency improvements. In addition, Europe is likely to experiment with traffic planning and continue the practice of securing of 'soft' voluntary pledges by the transport manufacturers for incremental efficiency improvements.

In both cases, the experience and policy focus of the different approaches should assist and aid each other. Incremental improvements in fuel efficiency standards from one country will eventually be universally adapted, and traffic management schemes tried in some places should assist those areas where such efforts have not been attempted.

7.4 Recent Developments in Automobile Emission Control Policy

As outlined above, there are a number of important trends in motor vehicle emission control policy that point towards potential next steps. One area of policy, emission control tailpipe standards, is constantly being amended. Section 7.5 below outlines the current differences in tailpipe standards and comments on the next potential steps for the EC and the UK. But as the trends above indicate, vehicle emission control policy has moved well beyond the tailpipe as the sole source of concern.

Discussed below are several areas of recent policy development associated primarily with passenger car air pollution. They cover the whole spectrum of questions facing the motor vehicle industry and policy makers. Current policy developments aimed at reducing sources of new vehicle emissions other than tailpipes are reviewed (in-use, I/M programs, refuelling & evaporative controls). New policy developments aimed at reducing emissions under certain conditions of vehicle use are reviewed (carbon monoxide emissions present during cold weather). Other policy developments aimed at improving the emission portfolio of the entire on-road fleet are discussed (accelerated scrappage and retrofit programmes). Finally, current policy developments that are widening the scope of traditional mobile source control policy are considered (off-road vehicles and mobile-stationary source trading). Policy development in the fuels area is discussed separately in Chapter 8.

After an explanation of the policy developments in these areas in various countries, some implications for future UK policy are introduced. Due to obvious limitations, some major developments in the control of pollution from freight vehicles and buses have not been sufficiently addressed. Yet some of the same trends for passenger cars are active in these areas also. Similarly, not all relevant vehicle emission performance regulations are reviewed below.

7.4.1 In-Use Emissions and Inspection and Maintenance Programs

In order to guarantee that vehicle emission control technology will maintain its optimal level of performance throughout the 'useful' life of the vehicle, the inspection of the emission control program and carrying out of necessary repair and maintenance is crucial. Various estimates of average in-use vehicles all indicate that they emit much more than their allocated emissions. Inspection and maintenance programmes are designed to improve in-use performance.

US estimates indicate approximately three to four times the allowed level of emissions are currently emitted by in-use vehicles on average. Other surveys have indicated that fewer than 10% of the vehicles in the US emit 50% of the road transport carbon monoxide. Proper identification of gross emitters and maintenance is imperative.

To combat this problem and to ensure that the investment in the emission control technology achieves maximum value, many different types of vehicle inspection programmes, or 'I/M' as they are called, exist. As the sources of vehicle and fuel emissions are progressively controlled, the value of good maintenance, expressed in very cost-effective emission control, is increased. The following is a review of various national inspection and maintenance programmes. The US programme is reviewed first because it is arguably the most advanced and the emission control and environmental benefits of the new programme are the most analyzed. The I/M programmes of various countries are outlined in Table 22 below.

Recent Developments in US I/M Programmes

As part of the US programme to help various areas of the country achieve ambient air quality levels, the 1990 US Clean Air Act proposed a federally-mandated I/M programme in many of the worst polluted areas in the US. In total, the programme is designed to achieve a fleet VOC reduction of 28% at the cost of $300/tonne of VOC removed. By contrast, the average stationary source VOC control measure considered in the new act was on average $5,000/tonne. In addition, it was estimated that some 15 million barrels of oil per year would be saved from the increased fuel efficiency from properly maintained vehicles.

Under the programme, areas that currently do not have any I/M programme will have to implement one, and areas with existing I/M programmes will have to implement an 'enhanced' programme.

The US 'basic programme' consists of a simple engine idle test for CO and HC that is required on a variety of periods according to the state where the vehicle is registered. Basic programmes can be implemented either at centralized state-run centres or at local garage and auto maintenance facilities.

The new 'enhanced programme' will be required in many of the worst ozone polluted areas in the country. The programme will include the following:

- testing at centralized facilities rather than at private garages;
- annual test of all 1968 and later model passenger cars and light duty trucks;
- steady-state testing for '68–'85 models;
- I/M 240 test plus and evaporative purge test '86 and later models (see below); and
- visual inspection for connection of catalyst and for leaded-fuel inlet restrictor presence.

The I/M '240 Test' will mimic the federal vehicle certification procedure and is about three times more accurate than existing tests. If successfully implemented, the test frequency may be reduced to once every two years rather than required annually.

The location of car testing is quite important to the programme's effectiveness rate. The EPA concluded that centralized programmes have the capacity to double the improvement of vehicle emissions performance compared to decentralized testing and repair. The trend in the US will be toward centralized testing and reporting of emission test results to local air quality programme managers to adjust the projected emission reductions and HC and NO_x control programmes.

> ### EC
> In August, 1992 a modest EC I/M directive (92/55/EEC) was adopted requiring member states to issue mandatory programmes. The stringency of such programmes is minimal compared to the new

US requirement. The major differences are the frequency of testing, the lack of a HC test, and the lack of a 'loaded' test.

The EC Directive requires member states to institute programmes that require all vehicles registered from 1994 to be tested after four years and then every other year. The minimum test is a idle CO check (0.5% and 0.3% when at 2000 rpm). 3-way catalyst equipped cars will be required to have a Lambda test to assure proper functioning of the closed loop system. The lack of prescriptive requirements in the directive will mean that member states will not be guarantied the extra benefit of higher performance I/M programmes unless they act independently to the degree allowed by the law.

UK

In-service emission tests for passenger cars and light duty vehicles in the UK were adopted in 1991 as follows:

1975 and later:	1200 ppm HC idle test
1975–1983:	maximum 6% CO idle test
1983 and later:	4.5% CO idle test
all vehicles:	no 'excessive' smoke
Diesel Vehicles:	no in-service inspection unlike application of roadside checks (UN ECE 24) in other countries.

An instrumented check for HGV and public service vehicles was introduced in September 1992, replacing the visual inspection method.

Some local authorities have established 'gross emitter' telephone call-in lines. Almost 80% of the over 1400 calls from April 1991 to March 1992 resulted in action being taken[79].

Germany

Regulations corresponding to the EC I/M regulation (92/55/EEC) were issued in Dec, 1992 to be applicable from December 1993. 3-way catalyst cars must have 0.5% at idle and 0.3% at high idle and lambda value is limited to 1 ± 0.03 unless specified differently by the manufacturer.

---

79 op cit note 15.

Table 22 In-Use Passenger Car Inspection and Maintenance Tests and Programmes

Country	Scope	Type	HC	CO (% vol)	Lambda	Age	Periodicity Frequency
EC	non 3-way	idle	none	type approval limit + 0.5%		3	1
	3-way	idle	none	0.5%	1±0.03	3	1
UK	'83 on	idle	1200 ppm	4.5%		3	1
	'75–'83	idle	none	6.0%		3	1
Denmark	pre-'84	idle	none	5.5%		5	owner change
	post-'84	idle	none	4.5%		5	owner change
	3-way	idle	none	0.5%		5	owner change
Finland proposed	pre-'87	idle	1000 ppm	6.0%		3	1
	post-'87	idle	600 ppm	5.0%		3	1
	3-way	idle + 2500 rpm	100 ppm	0.25%	0.97–1.03	3	1
US	basic (selected areas)	idle	220 ppm	1.2%		1	1
	enhanced (selected areas)	full load simulation	220 ppm	1.2%		1	2

Source: CONCAWE Report no. 2/92, 1992 and US EPA.

Remote Sensing of Vehicle Emissions

Another method of identifying the worst emitters in the fleet is to rely on remote sensing of vehicle emissions or random spot checking. Technology now exists in the form of infrared remote sensing systems (first developed at the University of Denver) to reliably identify gross emitters. If established at on-road locations, remote sensing can be an unobtrusive method of quickly screening numerous vehicles. Gross emitters identified through such a process could then be instructed to have their vehicle tested at permanent inspection and maintenance facilities, like those described above.

The value of such a system would be to efficiently identify the worst polluters and direct them towards corrective action. The value of a remote sensing screening mechanism would also be highest in those areas with the most infrequent I/M programmes. One study indicates that remote sensing can cost less than $0.50 per car compared to an average of $10 per car at a centralized I/M facility[80].

A remote sensor can also screen more vehicles than a permanent facility and would not inconvenience drivers to the degree that testing facilities can.

Implications and Future Directions for the UK

- Inspection and maintenance motor vehicle control programmes are very cost-effective relative to other incremental emission reduction policies. For example, in the US, centralized I/M programmes are expected to cost on average $5.70 per vehicle (1990$). This yields HC reduction benefits of approximately $300/ton. This compares to an average $3,000/ton HC control cost for implementation of the 1994 US tailpipe standards[81]. As road transport grows in the UK and as more of the fleet becomes controlled by catalysts, the benefit of insuring proper maintenance and operation of the emission control system will increase.
- The addition of a loaded test cycle to the current idle test in the UK will bring the I/M programme more in line with existing test cycle procedures which have recently been changed to acknowledge actual driving conditions.
- Centralized I/M test centres could be established on a pilot project basis in order to assess the difference in emission control benefits between government operated and private operated and licensed test facilities.
- Other test and repair arrangements should also be investigated. A proposed 'hybrid' version in California would combine several private independent test-only facilities with repair and testing taking place at privately qualified facilities.
- To locate catalyst malfunctions at an earlier date, the initial test for new cars should be advanced from the third year to the first or second year.

---

80 M Pitchford, 1992, *Remote Monitoring of Vehicle Emissions*, US EPA, Las Vegas, NV.
81 US Environmental Protection Agency 1991, *Ozone Nonattainment Analysis; Clean Air Amendments of 1990*, Washington, DC.

7.4.2 Refuelling and Evaporative Emission Control

Petrol vapours emitted during the *refuelling* of cars and refilling of petrol storage and distribution terminals and *evaporative* losses from cars are a significant source of total vehicle HC emissions. Some 15% of total VOC emissions in the EC are from these sources (10% from evaporative losses from cars, 3% from storage and distribution operations, and 2% from refuelling operations). Recent regulatory developments to address these sources of emissions are discussed below.

Petrol distribution and storage operations – Stage I

> EC
>
> Emissions from terminal and service station loading and unloading operations and petrol storage tanks, as well as loading and unloading emissions from barges and ships and road and rail tanker cars, are all subject to future control. The EC Environment Council recently adopted a Directive that includes the capture of unloading emissions at service stations – part of the so called 'Stage I' control measures.
>
> The Directive, when passed by the European Parliament, will exempt stations with an annual throughput of 100,000 litres or less. Those with a volume of 500,000 litres or less may apply to their Member state for a derogation. Some 4,700 service stations in the UK are candidates for derogation[82].
>
> As of the end of 1991 in the UK, 3.5% of the approximately 19,000 petrol stations had installed 'Stage I' vapour recovery equipment to capture HC emissions during the refilling of the station's main storage tanks[83]. Mandatory adoption of Stage I controls is opposed by the UK Petrol Retailers Association, representing independent service stations[84].

Auto Refuelling Vapour Control – 'Stage II' and 'On-Board'

During the refuelling of autos at service stations, petrol vapours escape to the atmosphere. Starting in certain ozone-polluted areas of the US in the early 1980's, these refuelling emissions were controlled by installing vapour collection devices at the end of gasoline pump nozzles. These devices capture the refuelling vapours displaced from the car's tank when new gas is pumped in

---

82 *ENDS Report 221*, 1993, p 36.
83 Institute of Petroleum data as reported in *ENDS Report 206*, 1992, London.
84 *ENDS Report 216*, 1993, p 37.

("Stage II"). They are then returned to the station's storage tank in the form of reclaimed fuel following condensation.

Another form of controlling these refuelling vapours is to install a carbon canister in the car itself which would capture the vapours during refuelling. Such an approach was originally championed by the EPA in the US, but auto company lobbying and testimony from the National Highway Safety Administration stalled the mandatory installation of such controls. This decision has been recently reversed in court however and it is now likely that on-board carbon canisters will be required in new car production in the US.

These so-called 'on-board' canisters, of a small capacity, are required in all new EC cars from January 1993.

Concerning 'stage II' pump vapour collectors in the EC, the proposal is still being analyzed and has not yet been formally proposed, but is likely to take place despite increased lobbying efforts by the refiners and distributors.

Other Countries

> **Germany** has passed regulations requiring the installation of Stage I and Stage II controls. As of January 1993, all new stations must have Stage II and existing stations will have three to five years to comply, depending on the volume of fuel sold.
>
> **Sweden** passed Stage I and Stage II regulations in January 1991 that require a phased-in application of the controls by January 1, 1995, which will cover 100% of all service stations.
>
> In **Switzerland**, all service stations must begin and complete by January 1995 the installation of Stage II controls. Controls must operate at a 90% efficiency level[85].
>
> Many cities and states in the **US** currently require the use of Stage I systems. Stage II system applications are going to be expanded as a result of the Clean Air Act of 1990.

Implications and Future Directions for the UK

- It is likely that Stage II vapour collectors will become mandatory in the UK at some point in the future – after the implementation of

---
85 op cit note 21.

Stage I measures. The relative cost effectiveness of the Stage II VOC control measure (compared with other stationary source VOC control measures), combined with the accompanying reduction of high-level toxic exposure scenarios, makes this decision a prudent policy choice.

- Like the US, the debate with the service station operators will revolve around the capital cost of the installation and the pay-back rate of the devices. At some point, due to the reclaimed vapours, the stage II devices reclaim the capital investment. This occurs sooner for high volume stations. The debate will therefore be similar to that currently occurring for Stage I implementation. Smaller stations will want exemption from the regulations or a longer phase in period.
- Analysis of the estimated capital expenditure and pay-back rates should take place now. Such analysis has been completed in the US and could be used as a guide.

Evaporative Emission Control

In addition to controlling refuelling and distribution operation emissions, vehicle manufacturers are also being required to control the evaporative emissions that result during a car's daily operation. Three sources of evaporative emissions are of concern. 'Running losses' occur as a vehicle is driven. 'Hot soak losses' occur after a fully warmed-up vehicle is left to stand. 'Diurnal' temperature emissions occur as the car warms and cools due to daily changes in the ambient temperature.

Various test methods are employed to measure these sources of emissions and various control devices are used by manufacturers to control such emissions. The major future area for policy development is in the refinement and standardization of the test methods. The main way to address evaporative emissions is to include them into the vehicle test cycles and certification programmes. Both the EC and the US include evaporative test procedures in current test cycles.

Implications and Future Directions for the UK

- The major difference between the EC and US evaporative emissions test is the inclusion of 'running losses' in the US and the exclusion of running losses in the EC. If there would be any development in EC and, hence, UK test methods, it most likely would be a move to incorporate running losses into the test.
- Planned action to reduce the permitted volatility of petrol sold in the

summertime in the UK will reduce evaporative emissions from the entire fleet[86].

7.4.3 Carbon Monoxide Ambient Problems: Cold-temperature Emission Standards and Oxygenated Fuels

Recognizing that higher CO emission levels are created from vehicles operating at relatively colder ambient temperatures and that such vehicles have difficulty in meeting CO vehicle emission standards set with more ordinary ambient temperatures in mind, the US Clean Air Act has implemented a carbon monoxide programme embracing two elements. The first is a change in the test procedure for new vehicle certification and the second is the mandatory addition of oxygenates to petrol in certain areas suffering from above-standard carbon monoxide levels.

Cold-Temp CO Test Standard

Beginning in 1994 with 40% of the new vehicle sales volume and increasing to 100% of the new vehicle sales in 1996, cars and trucks with a GVW of 3750 lbs will have to emit ten grams CO/mile (12.5 grams/mile for light-duty trucks over 3750 lbs). These standards are based on testing the vehicles at 20 degrees F (−7 degrees C). EPA is seeking comments on a proposal to allow compliance to be averaged in some way across the new vehicle fleet.

Expected Effect

The EPA estimates mobile source CO emissions, when measured at 20 degrees F, will be reduced by approximately 20–29% from the revised test procedure. When averaged over all ambient operating temperatures, the control of CO emissions under cold regimes will result in annual CO emission reductions of between 2.6 and 3.1 million tonnes by the year 2000 and 5.8 and 7.7 million tonnes after complete fleet turnover.

Oxygenated Fuels for High CO Areas

To reduce ambient levels of CO, those areas which exceed the national ambient standard must add oxygenates (methanol or ethanol) to gasoline to a 2.7% mass content. This requirement is for the colder temperature winter fuel sales only. During the first season when it was applied (the 1991–1992 winter), some 41 cities in the US, representing 31% of total US gasoline sales volume, were required to add oxygenates.

86 op cit note 15.

Preliminary results of the programme from the first year have indicted dramatic success. Based on using the new mandated oxygenated fuels, 38 urban areas reported only two days where carbon monoxide levels were above the ambient air quality standard. This compares with 43 days exceeding the standard during the previous year. California went the entire winter without one day violating the federal standard and Anchorage, Alaska had its first violation-free winter in 20 years.

Implications and Future Directions for the UK

- CO levels in the UK are likely to continue to be above the EC and WHO directives and guidance levels on occasion. Because CO levels are elevated at colder times of the year, a cold temperature CO standard could be useful.
- EC member states with colder average ambient conditions than the UK may benefit in greater proportion. Preliminary results from the US indicate the enormous potential of oxygenated fuels for reduction of vehicle-related CO emissions.
- Fuels-based emission control policies may be more appropriate for particular climates and regions that are experiencing air quality programmes that other areas are not. The ability of fuels-based programmes to address niche markets that vehicle-based programmes cannot should be investigated further.
- The addition of an ambient cold-temperature test condition to the vehicle performance test cycle would ensure that vehicles are designed with all the appropriate operating conditions in mind.

7.4.4 Accelerated Scrappage Programmes

Developments and applications of vehicle emission control technology, such as catalytic converters, are quite successful in reducing per-vehicle pollution from new vehicles. The programme's total efficacy is, however, partially determined by that proportion of the fleet that is equipped with the technology, the part that is unequipped, and the rate of fleet turnover.

The matter is further complicated as the incremental cost of emission control technology itself reduces the demand for new vehicles. This acts to slow down fleet turnover and delay the achievement of wider ambient air quality benefits. Some have argued that the implementation of emission control programmes in the United States in the 1970s, at a time when the cost and technical capability of the emission control technology was not as developed as it could have been, reduced vehicle fleet turnover. This exasperated the pollution problem by keeping old 'dirty' vehicles on the road past their normal retirement or 'scrappage' time period. Although this effect may have taken place for a limited time period, it is unlikely to have any long term significance as the differential between a slower fleet turnover and a relatively less-controlled fleet widens.

While the cost of proposed changes to vehicle design must always be considered in determining the potential effectiveness of any air pollution control programme, the potential retardation of fleet turnover rates due to increased vehicle price should not be used as an excuse for avoiding technological changes. Rather, it should serve as an analytic tool and forcing mechanism to develop the most cost-effective control regime, and to implement programmes in recognition of and compatible with the economic cycles of the motor vehicle industry and consumer spending habits.

In addition, recognition of variable price elasticities of demand for vehicles at opposite ends of the cost spectrum can also allow for the differential application of fiscal incentive and taxing programmes to have the maximum effect.

The problem of fleet turnover, or the persistence of older more polluting and less efficient vehicles on the road, has resulted in several unique policy experiments. Programmes that encourage the accelerated retirement or scrappage of older vehicles, incentive schemes to turn in old cars and buy new ones, and programmes to retrofit emission control technology to existing vehicles are reviewed below.

Accelerated Vehicle Scrappage Programme Background

The idea of removing older, more polluting vehicles from the road and replacing them with newer, cleaner and more efficient ones makes intuitive sense. In the US, relatively high old-car per-mile emissions and their continuing use indicates that, although pre-1971 cars in the US drive only 1.7% of all national vehicle miles travelled, they are responsible for almost 5% of the national NO_x, 7% of HC, and 7.5% of CO[87]. If these older cars were retired earlier than when they would have been, then reductions in emissions of air pollutants increases in passenger safety and improved energy efficiency could be achieved. In addition, the stimulation of new vehicle purchase could be an important economic benefit.

---

87 W L Schroeer, 1991, *Accelerated Retirement*, EPA information document, Draft, US Environmental Protection Agency, Washington, DC.

As vehicle emission control technology improves, the air pollution difference between new and old cars increases (see Figure 22 below).

Figure 22: HC emissions from average model - year car (g/mile)

Source: Accelerated Retirement, EPA Information Document Draft, October 2 1991, WL Schroeer, US Environmental Projection Agency, Washington DC

Estimating the overall effect on air pollution of scrapping older cars is difficult. It depends on several variables including the amount of miles driven by older -v- newer cars, the period of potential use past the scrappage date of the older vehicles, the percentage of older to newer vehicles in the fleet (or in regional fleets) and the estimation of the improvement of an accelerated retirement programme over and above the benefit of natural retirement of vehicles. Every year, owners in the US scrap 20% or more of all 16 year-old vehicles, yet some portion of older vehicles will always remain in service.

It is important to know the number of miles driven on average by younger cars compared with older cars. New cars drive many more miles than old ones, so scrappage programmes must only credit reduced pollution for the miles avoided. This benefit is limited in time as all cars are eventually scrapped. Figure 23 below portrays the distribution of miles driven here in the UK according to vehicle age and motor size.

Figure 23: Vehicle Age and Mileage by Engine Size[88]

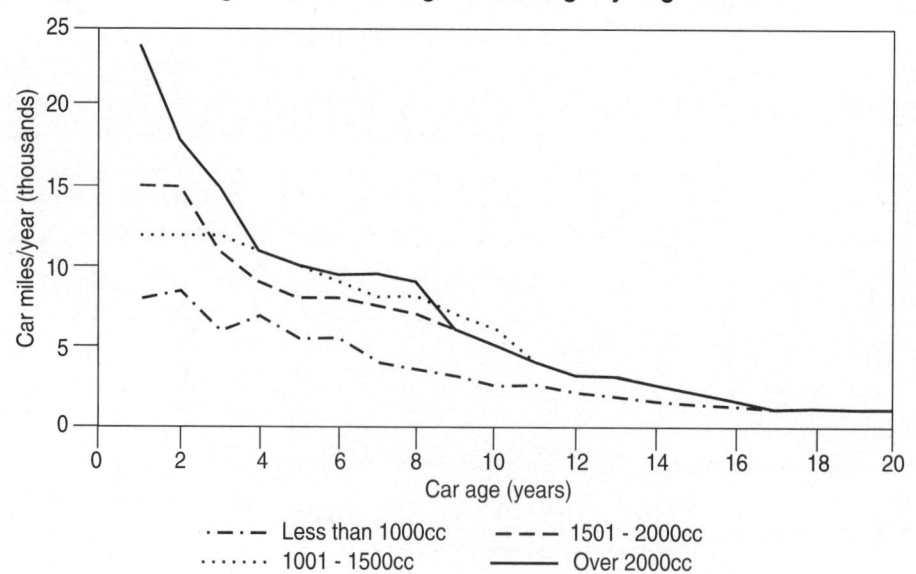

An appropriate vehicle scrappage programme must address the following questions:

> Which vehicles are eligible (eg, registered in appropriate area, operable, licensed, of certain age)?
>
> What are the emission reduction benefits of scrappage?
>
> What fiscal incentive, if any, is appropriate and how should this be provided (eg, in the form of a coupon for new car purchase, a discount, in cash)?
>
> How should emission reduction credits be calculated and what is the period that such emission reduction benefits are viable?

These questions are especially important if the environmental benefits of scrappage programmes are going to be used to delay or offset environmental improvements or implementation of other types of standards.

The level of emission credit for each car and type of vehicle must be determined. The programme must be designed and implemented in a way that roadworthy cars that are actually driven or have the potential to be driven are retired. Cars should also be licensed and insured. The programme may only be able to be implemented on a one-time or periodic basis to avoid 'hoarding' of

---

88 D Martin & L Michaelis, 1988, "National Travel Survey, 1985/86", Department of Transport, HMSO, London, 1988; as reported in *Road Transport and the Environment*, Policy, technology and market forces, Financial Times Business Information, 1992, London.

old polluting cars. Vehicles that are scrapped must be destroyed so that they do not re-enter the market.

An Example: The UNOCAL SCRAP Programme

In 1990, UNOCAL Corporation of California carried out an accelerated scrappage programme offering $700 for pre-1971 models that met certain requirements. Offers to sell totalled more than UNOCAL could pay for and they eventually bought and retired 8,350 vehicles. This programme has been analyzed in detail. However, the interpretation of the results of the programme apply only to the southern California area where the program was implemented and are not transferable. Some elements critical to the design of successful scrappage programmes were not tested in this experimental programme (such as duration and periodicity of programme).

Unocal reported the following results:

- Total fleet mileage after scrappage and with replacement was approximately the same.
- Median age of scrapped car – 1981
- Median age of replacement car – 1983
- *Emission Reduction*   *Conservative Estimate*   *Liberal Estimate*
  HC   663 tons/yr   717 tons/yr
  CO   3,008 tons/yr   3,481 tons/yr
  NO_x 27 tons/yr   75 tons/yr
  PM   52 tons/yr   63 tons/yr
- Price Required for Incentive to Scrap
  The $700 offered by UNOCAL was in excess of the amount necessary to entice sellers according to 'blue book' prices and conditions of the cars scrapped.

The biggest variable in determining the benefits of any scrappage programme is estimating the amount of mileage the replacement cars will drive.

In a separate assessment of vehicle scrappage programs, the US Office of Technology Assessment concluded that well targeted and designed programmes could achieve environmental benefits at costs equal to or lower than those of other emission reduction options. The particular emission reduction benefits of various programmes depends on the surrounding vehicle control programme. The benefits of such scrappage programmes are likely to be reduced in the future as other programmes advance in sophistication and breadth.

Implications and Future Directions for the UK

- Accelerated scrappage programmes offer a unique market-based solution to localized air pollution programmes.
- These scrappage programmes are being implemented in geographic areas that have specified emission reduction targets.
- In the UK, where local regions do not have localized ambient air quality goals, this motive force is missing. Although nation-wide air quality goals in the UK can provide the framework with which to evaluate such programs, a formal recognition of the relationship between local air quality improvements and national goals is still needed. Accelerated scrappage programmes could have an impact on the local economy and potentially the environment. Implementing a pilot programme would create a demand for much needed vehicle fleet and driving condition information.

7.4.5 Mobile Source Retrofit Programmes

The other method that is being used to address the problem of older polluting vehicles is to retrofit such vehicles with emission control technology or outfit them to operate on cleaner burning fuels. Four programmes are discussed below.

(i) **US – Urban Bus Retrofit Programme**
Under the US Clean Air Act of 1990, the EPA must issue standards for the retrofit and improved maintenance of urban buses that have their engines replaced or rebuilt from 1995 on. According to proposals, urban transit authorities would have two options. They could rebuild the engines (incorporating the latest diesel trap oxidizer designs) to meet a particulate standard (most likely of 0.1 g/bhp). The second option would be to use a combination strategy of early vehicle retirement, trap retrofits, and/or the use of alternative fuels, such as compressed or liquefied natural gas, so long as the aggregate emissions were equal to that under the first option.

(ii) **California Bus Methanol Retrofit**
Preliminary results of retrofitting in-use transit buses to run on methanol are encouraging. This programme in Southern California is intended to explore the economic feasibility as well as environmental benefits of bus conversion. The retrofits involve adapting existing diesel-fuelled engines to burn methanol.

The retrofitted buses emit 75% less particulate matter and 50% less NO$_x$ and slightly increased HC and slightly decreased CO. The cost for conversion at small-scale production was $22,000/bus. This can be reduced to $12,000 with larger volume production. Operating costs are also higher due to the use of 3% concentration of the fuel-additive Avocet, a proprietary cetane improver[89].

It should be noted that this program is just one of many bus conversion programmes already in operation. A wider review is necessary before conclusive environmental and economic parameters can be assessed.

(iii) Athens Greece Programme

To address persistent transport pollution, future projections of increased vehicle traffic and pollution, and an average fleet age of 11 years (twice the EC average), a new set of measures was proposed in November 1990 that included both vehicle scrappage and retrofit[90]. The programme was meant to reduce the fleet turnover rate from the approximate 25 years to levels seen in other countries (Germany is 15 years, for example). The programme would include the following elements:

> A 50–60% reduction in new 3-way catalyst equipped vehicle purchase tax and a five year exemption of the annual road tax when accompanied by the scrappage of an older car.

> In-use cars retrofitted with catalysts will be exempt from road tax for two years. One additional year of road tax exemption would be granted if retrofitted with a carbon canister.

> Prohibition of polluting vehicles from a metro Athens area.

Projected Potential Results:

At the time of the analysis, if the scrappage programme had been adopted and 290,000 light duty cars and 34,000 light duty trucks were replaced over a 2 year period, CO in greater Athens was expected to be reduced by 35% and VOCS by 20%.

---

89 South Coast Air Quality Management District, 1990, *Annual Report*, Technology Advancement Office, Los Angeles, California.

90 Z C Samaras and K N Pattas, 1992, "Assessment of an Optimized Policy for Reduction of Road Traffic Emissions in Athens", Aristotle University of Thessaloniki, Greece in *The Motor Vehicle and The Environment – Demands of the Nineties and Beyond*, Int'l Symposium on Automotive Technology and Automation, published by Automotive Automation Ltd, London, 123–130.

(iv) **Santiago Chile Passenger Car Retrofit Programme**

Santiago Chile is plagued by high CO, O_3 and PM levels. A pilot programme launched by the government in 1991 retrofitted 28 taxi cabs with oxidizing catalysts[91]. This well controlled experiment concluded that:

- On average CO was reduced 55%–65%.
- On average HC was reduced 70%–80%.
- If applied to all taxis in Santiago, then a 12% ambient CO reduction and a 15% ambient HC reduction could be expected.
- The average retrofit cost was $275 ($1991).

Implications and Future Directions for the UK

- Like scrappage programmes, retrofit programmes are perfect for creating local solutions to local problems. Unlike scrappage programmes, the benefits of vehicle retrofit are likely to be greater and the programme easier to promote and analyze.
- Many parts of the UK fleet in various towns and areas could benefit from retrofit programmes. Bus and cab fleets are the most likely candidates. Missing again is the local air quality improvement imperative that exists in the US. But where improvement of urban air quality is an objective of local government, as it is in many cities in the UK, retrofit programmes, combined with selective access to cleaner vehicles (like the proposed Athens programme), could be quite attractive.
- Further bus privatisation efforts around the country could, as a condition for sale and route access, require improved environmental fleet performance.

7.4.6 Off-Road Vehicle Emission Policy Development

As pollution from HGVs and passenger cars is reduced, attention is shifting to a relatively unaddressed part of the mobile source emission inventory – tractors and construction equipment used in agriculture, road construction, forestry and other practices. Regulatory developments are discussed below.

---

91 G Reinke, A Saez, R Katz, 1992, "Converters Retrofitted to Carborator Vehicles", in *The Motor Vehicle and The Environment – Demands of the Nineties and Beyond*, Int'l Symposium on Automotive Technology and Automation, published by Automotive Automation Ltd, London, 209–214.

US and California

Both the US and California have been formulating emission limits for off-road diesel vehicles. California has taken the lead by issuing standards in early 1992. The federal government has agreed to adopt similar measures for 1996, but will most likely not adopt the more stringent second phase 2001 standards.

In 1992, California adopted emission standards for off-road diesel engines as the following table indicates.

Table 23 California Off-Road Diesel Vehicle Emission Standards

	NO_x	PM
1996 Diesel > 175HP	6.9 g/BHp-hr	0.4 g/BHp/hr
2000 Diesel > 750HP	6.9 g/BHp-hr	0.4 g/BHp-hr
2001 Diesel > 175HP	5.8 g/BHp-hr	0.16g/BHp-hr

EC and UNECE

Discussions are taking place between the EC and the UNECE to develop emission standards for HC, NO_x, CO and particulates for off-road vehicles. The joint regulatory development path being used by the EC and the UNECE is in itself an improvement. Previously, regulations for passenger car and HGVs have been developed by one organization and adopted by the other at some later time. In addition, the goal of these discussions is to achieve harmonization with the regulatory developments taking place in the US.

Implications and Future Directions for the UK

- The prevalence of off-road vehicle emissions in the UK inventory will increase over time. Policies to address these sources should be welcomed.

7.4.7 Mobile -v- Stationary Source Emission Trading

With the advent of the acid rain emissions trading programme in the US and increased interest in market incentive-based environmental regulations in general, it was just a matter of time before a proposal to trade reductions of mobile source air pollution emissions with stationary source emissions was made.

After a year of investigation, a proposal has been agreed by the South Coast Air Quality Management District of Southern California to develop such a programme. The RECLAIM programme will initially allow for the trading of excess emission reduction credits between stationary sources. Similar to the acid rain emission credit trading programme, sources that control more than their required amount would be allowed to generate emission trading credits, which could be purchased by other businesses in lieu of reducing their own emissions by the required amount.

If successful, there is discussion of expanding the programme to incorporate fleet emission reductions and other mobile source emission reductions.

Implications and Future Directions for the UK

- Without local air quality management plans that require a set amount of emission reductions, the utility of emissions trading programmes for the purpose of local and national pollution reduction is limited in the UK.
- When an optimal emission reduction programme is required for whatever reason, then emission credit and trading schemes bear investigation. The potential economic benefit of such programmes is great. It is estimated that the new acid rain emission trading programme in the US will save approximately $1 billion or 25% of the cost of a non-emission trading system. Perhaps the greatest benefit of any emissions trading scheme is the incentive it gives for cost-effective technology development.
- If this or other programmes were to be implemented in the UK, it is imperative that the current portfolio of emission information is improved. Most environmental policies employing market mechanisms or fiscal incentives (especially emissions trading schemes) are information intensive.

7.5 Passenger Car Tailpipe Emission Control Measures

Introduction

Passenger car tailpipe emission standards for HC, NO_x, and CO for the US, California, the EC, and the UK are given in Tables 24 to 27 below. Also listed are combined HC + NO_x standards. For each table, the US and California standards have been converted from grams/mile as determined by the US 'Federal Test Procedure', to grams/kilometre as determined by the EC Urban test cycle. For the combined HC + NO_x standards, hypothetical combined standards have been supplied for the US and California. The same procedure has been followed for early EC standards prior to the Consolidated Emissions

Directive. The historical development of EC passenger car emission standards is given in Table 28.

Following the tables, the implications of the differences and potential future developments is discussed.

Table 24 HC & NO$_x$ Combined Passenger Car Emission Standards[1] (g/km EC test cycle)

California[2]		US		EC	
Year	Limit (g/km)	Year	Limit (g/km)	Year	Limit (g/km)
Pre-control	15.3	Pre-control	15.3		
'70–'71	7.5	'70–'71	7.5		
'72–'73	4.3	'72	6.7	Pre-1977 no NO$_x$ control	
'74	3.4	'73–'74	4.4		
'75–'76	2.1	'75–'76	3.3		
'77–'79	1.4	'77–'79	2.5	'77–'78	4.2–6.7
'80–'81	1.0	'80	1.7	'79–'83	3.6–5.7
'82–'92	0.57	'81–'93	1.0	'84–'91	4.7–6.9
'93	0.47			'92–'95	0.97
'94–'99	0.38 TLEV 0.25 LEV	'94–'03	0.48	'96–'02	0.5
'00–'03	0.17 ULEV 0 ZEV	'04	0.24		

1 For purposes of comparison, separate HC and NO$_x$ emission standards have been added together for all years shown for the US and California and for those years prior to the EC Consolidated Emissions Directive 91/441/EEC (prior to 1984).
2 Conversion factor for US grams/mile Federal Test Procedure to EC Test Cycle grams/km is 0.72 per C Holman, J Wade & M Fergusson, January 1993, "Future Emissions from 1990–2050: The Importance of the Cold Start Penalty: A Report for World Wide Fund for Nature UK, Earth Resources Research, London.

Table 25 Carbon Monoxide Passenger Car Emission Standards (g/km EC test cycle)

US[1]		California[1]		EC		UK	
Year	Limit (g/km)	Year	Limit (g/km)	Year	Limit (g/km)	Year	Limit (g/km)
pre-control	81	pre-control	81	–		–	
'70–'71	30.6		30.6	'70–'73	25–55		
'72–'74	25.2	'72–'74	25.2			'73–'76	25–55
'75–'79	13.5	'75–'79	8.1	'74–'78	20–44	'77–'82	20–44
'80	6.3	'80	8.1	'79–'83	16–36	'83–'85	16–36
'81–'93	3.06	'81–'92	6.3	'84–'91	14.27	'86–'92	14.27
		'93–'99	3.06	'92–'95	2.72	'93–'95	2.72
'94–'03	3.06	'99–'03	1.53 ULEV / 0 ZEV	'96–'02	2.2	'96–'02	2.2
2004–	1.53						

1 Conversion factor for US gm/mile Federal Test Procedure to EC Test Cycle grams/km is 0.90 per Holman Wade & Fergusson 1993.

Table 26 HC Passenger Car Emission Standards (g/km EC test cycle)

US[1]		California[1]		EC		UK	
pre-control	10.8	pre-control	10.8	–		–	
'70–'71	3.0	'70–'71	3.0	'70–'73	2–3.2		
'72–'74		'72–'74				'73–'76	2.32
'75–'79	1.1	'75–'79	0.65	'74–'78	1.7–2.7	'77–'82	1.7–2.7
		'77–'79	0.3				
				'79–'83	1.5–2.4	'83–'84	1.5–2.4
'80–'93	0.3	'80–'93	0.28				
'94–'03	0.18	'94–'03	0.18	[HC + NO_x combined as of 1984]			
2004–	0.09						

1 Conversion factor for US grams/mile Federal Test Procedure to EC Test Cycle grams/km is 0.72 per Holman Wade & Fergusson 1993.

Table 27 NO_x Passenger Car Emission Standards (g/km EC test cycle)

	US[1]		California[1]		EEC		UK	
pre-control	4.5	pre-control	6.2	–		–	
'73–'76	2.2	'72–'76	2.3				
'77–'80	1.4	'77–'79	1.08				
'81–'93	0.7	'80–'81	0.7	'77–'78	2.5–4.0	'82	2.5–4.0
'94–'03	0.3	'82–'93	0.3	'79–'83	2.1–3.4	'83–'84	2.1–3.4
'04–	0.15	'94–'99	0.29 TLEV				
			0.14 LEV	[HC + NO_x combined as of 1984]			
		'04–	0.14 ULEV				
			0 ZEV				

1 Conversion factor for US grams/mile Federal Test Procedure to EC Test Cycle grams/km is 0.72 per Holman Wade & Fergusson 1993.

Table 28 EC Passenger Car Emission Standards (gm/km)

Directive	EEC Date	HC	NO_x	HC+NO_x	CO	PM	Year Adopted by UK
ECE15; 70/220/EEC	1970	2–3.2	–	–	25–55		1973
ECE 15/01; 74/290/EEC	1974	1.7–2.7	–	–	20–44		1977
ECE 15/02; 77/102/EEC	1977	1.7–2.7	2.5–4	–	20–44		1982
ECE 15/03; 78/665/EEC	Oct '79	1.5–2.4	2.1–3.4	–	16–36		1983
ECE 15/04; 83/351/EEC	Oct '84/'86			4.7–6.9[1]	14–27		1985
ECE 83; 88/76/EEC	1988	Not implemented due to anticipation of Consolidated Directive					
ECE; 91/441/EEC	'92–'95			0.97	2.72	0.14	1993
MVEG 1991 proposal	'96–'02			0.5	2.2	0.08	–

1 Type Approval Standards

Discussion

The most obvious interpretations of the development of passenger car tailpipe standards outlined above are: (1) they have been getting progressively tighter over time, and (2) the general trend is for standards to begin in California, then to be adopted in the US, then the European Community and UNECE, and finally (for early EC Directives when member state implementation was optional) for the measure to be implemented in the UK. This trend in California leadership not only covers the past, but can also be projected into the future. California's program of ultra-low emission vehicles and zero emission vehicles will again set the standard for emission control performance. Further observations follow.

7.5.1 The 1994 US standards and the Proposed 1996 EC standards

With the proposed 1996 EC standards of 0.5 g/km HC+NO$_x$, 2.2g/km CO, and 0.08 g/km, the EC has implemented tailpipe emission standards on a par with the next phase of the US 1994 standards. The major differences that exist between the two sets of standards are:

- The US will maintain a two year lead in implementing the next phase of emission standards.
- The EC maintains a combined HC + NO$_x$ standard. These standards continue to be separate in the US.
- The EC has a more stringent CO emission standard than the US – 2.2 g/km compared to 3.06 g/km.
- Particulate matter standards for diesel vehicles in the US and California remain more stringent than those for the EC – 0.06 g/km US to 0.08 EC.

7.5.2 The 1994 California standards and the Proposed 1996 EC standards

When compared to the 1994 California standards, the 1996 proposed EC standards can be put into proper technological perspective. The '94 California HC and NO$_x$ standards are significantly more stringent than the 1996 EC standards. Perhaps more significant than the differences in the numerical emission standards, however, are other refinements in the California programme of control.

- The California transitional low emission standards (TLEV) for an arithmetically combined HC+NO$_x$ standard is 0.38 g/km compared to 0.50 g/km for the EC.
- This difference is even greater when comparing the '94 California low emission vehicle standard (LEV) to the proposed EC standard 0.25 g/km to 0.50 g/km.
- The California standards are likely to be adopted by other US states, bringing the US market share for the California-standard cars to approximately 50% of the market.

- The California programme introduces an important 'fleet averaging' method of compliance with the HC standard (non methane organic gases). 80% of the fleet must meet the numerical standard on an individual vehicle basis, but the rest of the fleet can comply by having their emissions averaged into the overall fleet. The fleet-averaged standard is then reduced on an annual basis. This will allow much more flexibility on the part of manufacturers in terms of compliance and the manufacture and introduction of new models.

- The California and US programmes have moved beyond an unspecified HC standard to specify either non-methane hydrocarbons (US) or non-methane organic gases. Both of these more specific chemical representations of hydrocarbon standards are meant to focus control efforts on the reactive portion of the hydrocarbon emission stream.

- The California standards refine this principal one step further by allowing for an adjusted emission rate, depending on the reactivity of the non-methane organics emitted. For example, if vehicles are fuelled with M85 (85% methanol), compressed natural gas or liquefied petroleum gas, they will have a reduced emission limit according to the reactivity of the emissions produced from vehicles using these fuels. The California Air Resources Board will determine the adjustment factors for various fuels.

- Both of these methods reward emission control strategies that lower the most reactive component of vehicle exhaust. They also create an incentive for introducing fuels that will reduce ozone formation.

- Another major development in the California system is the development of an emission credit scheme similar to that implemented with US fuel economy standards. For manufacturers that have a sales-weighted emission average lower than the allowable fleet average, emission credits will be granted that can be used against future years. These credits will be marketable between manufacturers to some degree. This scheme thus rewards early compliance.

- Both the '94 California standards and future EC standards may require the use of the electrically-heated catalyst to reduce the pollutants emitted during the first few minutes of vehicle operation before the catalyst has had time to warm up. Since the advent of electrically-heated catalysts (EHC) in 1990, their performance and cost has been dramatically reduced. In 1990 the EHC required a 20 second pre-start and 15 second post-start heat-up phase in order to become operable. This has been reduced to a 15 second post-start heat-up phase. The power requirements have been reduced from 600 to 170 amps. The energy requirements reduced from 124 to 15 watt-hours and the projected costs from $170 to $65[92].

7.5.3 The Next Generation of Standards

Both the current California law and the new US Clean Air Act achieved an important milestone in terms of emission reduction planning. Not only did they succeed in updating existing tailpipe standards, but they were successful in laying out the generation of standards beyond the new updates.

The Clean Air Act includes so-called 'Tier II' standards scheduled to be implemented beginning in 2004 (if necessary) and the California law continues the TLEV and LEV programme begun in 1994 by implementing more stringent standards beginning in 1999 – ultra-low emission vehicles (ULEV) and zero emission vehicles (ZEV).

For the first time, auto manufacturers have clearly defined numerical emission targets set far in advance to allow for necessary design time. In the case of the federal standards, a full 15 year lead time is provided for the 2004 standards and, even then, they will be phased in over several years.

Other notable developments in the 2004 US standards or the 1999 California standards include:

US 2004 standards

- A further 50% reduction in the HC, NO_x, and CO passenger car standards from the 1994 emission standards. Yet as drastic as these reductions appear, they will be technically overshadowed by the requirements of the California standards which will take on an increasingly large share of the US market. For example, the US 2004 HC standard will have already been in place in California cars for ten years, the NO_x standard for eight, and the CO standard for four years.

California 1998 and beyond

- As mentioned above, these standards will set the pace for passenger car tailpipe emission controls.
- The most notable development is the requirement in California (and those states that adopt the California programme) for 'zero emitting

---

[92] M J Bradley, 1993, *US Motor Vehicle Emissions Update*, North East States Coordinated Air Use Management, Boston.

vehicles'. By 1998, 2% of manufacturers sales must comply with zero emissions of HC, NO_x, CO, and PM. This percentage is increased to 10% of new car sales by the year 2003.

Implications and Future Directions for the UK and EC

To the degree that developments in the US and California illustrate the technical potential of passenger car emission reduction and to the degree that the environmental situation requires such actions in the EC and the UK, the following future directions can be gleaned from US developments. The influence of US developments on EC policy must be considered in the context of on-going EC mobile source policy discussions. Those discussions taking place in the Motor Vehicle Emissions Group are addressing many of the developments noted above. Some US developments that are not on the formal agenda of that group are mentioned below together with items that are subject of discussion.

Stringency

- It is clear that the 1996 EC standards should be viewed as interim ones and that more stringent targets will be developed shortly. Given the trend toward universal standard stringency (and technical responses), the developments in the US should set the target for the next and subsequent rounds of EC standards.
- It is likely that EC and UK manufacturers will not be granted the lead time that the new US developments have granted US manufacturers and importers.
- A 'zero emitting vehicle' programme, either implemented as part of urban pollution control strategies or as part of fleet programmes, will most likely be implemented in some EC country within the near future to the degree that EC law allows. The Commission and member states should fund suitable pilot projects in this area or should seek to encourage participation in the new US market.

Emission Averaging and Standard Phase-in

- It is likely that as standards become more stringent, they will be phased in over time in terms of the proportion of the fleet that has to comply. It is also likely that some sort of fleet-weighted emission average approach, as has been implemented in the new California standards, will be developed in the EC.

Marketable Emission Credits

- It is unlikely that a system of trading emission reduction credits between mobile and stationary sources or between manufacturers of mobile sources will be developed in the UK or EC without the development of more localized air pollution control targets and plans. The latter development is a necessary one, while the former is a way to optimise control strategies.
- Such integration could take place in the area of VOC control strategies, where mobile and stationary source emissions are being considered together.

Focussed Photochemical Oxidant Policy

- It is likely that further HC and NO_x emission control policies (which are implemented in part to reduce photochemical smog) will be based on a refined acknowledgement of the photochemical oxidation creation potential of various hydrocarbons. It is likely that the next round of EC standards could move to non-methane hydrocarbons or organic gases, rather than to total hydrocarbon emissions.
- The recognition of the contribution of various types of fuels towards certain environmental problems will improve dramatically over the next few years. This knowledge will affect transport environmental policy. The CO_2 generating potential of various fuels is one such example. The investigation of alternative fuels for their impact on HC, NO_x and CO emissions is another.
- It is likely that HC and NO_x emission standards in the EC will be separate at some point in the future as this allows for better optimization of emission control strategies in recognition of varied environmental goals.

7.6 Vehicle Speed and Emission Control

The speed of motor vehicles has a great impact on the amount of air and noise emissions produced and the amount of fuel used. The relationship between speed and these variables is increasingly becoming defined. As this happens, it will be possible to evaluate the effect of speed regulations and implement such measures where and when desirable. The relationship between speed and noise is discussed in Volume 3 of the Foresight project.

The relationship between driving behaviour, vehicle speed, and test cycles is important to note. Recently, the European test cycle was changed to acknowledge the increased emissions at higher speeds and to provide a better representation of driving behaviour. The US is also looking at their current passenger car Federal Test Proceedure to make it more accurately reflect actual driving conditions and the emissions of vehicles at higher speeds. Preliminary work indicates two problems: (1) higher than predicted NO_x emissions during 60 miles per hour 'cruise' conditions, and (2) higher CO and HC emissions during enriched driving phases such as acceleration and high speed[93].

A recent study on speed and emissions is outlined below.

Vehicle Emissions and Speed

Recent work has analyzed the relationship between vehicle speed and emissions of NO_x, CO, HC, and CO_2. Concerning CO_2 and NO_x, Fergusson and Holman have estimated the relationship between fuel consumption and speed for a variety of vehicle types[94]. The results are shown in Table 29 below.

Table 29 Emissions Benefits from Controlling Car Speeds

Speed	Carbon Dioxide Total	Emissions Saved	% Saved
Current regime	73315	–	–
Enforced 70 mph	71063	2253	3.1%
Enforced 60 mph	69522	3793	5.2%
Enforced 50 mph	68155	5160	7.0%

Speed	Nitrogen oxides Total	Emissions Saved	% Saved
Current regime	819.8	–	–
Enforced 70 mph	787.3	32.5	4.0%
Enforced 60 mph	762.3	57.5	7.0%
Enforced 50 mph	724.2	95.5	11.7%

Notes: All emissions data are in thousands of tonnes (kt)
Calculations are for 1992 total emissions (estimated)

Source: The effect of vehicle speeds on emissions, an update using 1991 speed data, WWF UK, August 1992

93 op cit note 6.
94 World Wide Fund for Nature UK, 1992, *The effect of vehicle speeds on emissions*, Godalming.

The implications of this relationship in terms of fuel economy and reduced CO_2 emissions is significant. By reducing average motor way speed from 115 kph to 110 kph (or strict enforcement of the 70 mph motorway speed limit), a total car fuel consumption savings of 2.4% would be achieved. This is increased to 4.4% and 5.8% for an enforced speed limit of 60 mph and 50 mph respectively. The NO_x reductions from various lower speeds are also depicted. Enforcement of the current 70 mph limit on motorways carries an immediate benefit of 4% lower NO_x reductions. This is due to the amount of road traffic speed above the legal limit. This is depicted in Table 30.

Table 30 Current Road Use and Car Speed Characteristics

Road Type	Proportion of Traffic	Speed in kph Average	Standard Deviation
Motorway	14.1%	116	17.0
Urban Road	45.6%	40	7.0
Single Carriageway	36.3%	77	17.5
Dual Carriageway	4.0%	108	18.0

Notes: Based on Department of Transport (1992)
Standard Deviation is a measure of the variation of speeds

Source: The effect of vehicle speeds on emissions an update using 1991 speed data, WWF UK, August 1992

Implications and Future Directions for the UK

- Reducing motorway vehicle speed meets multiple policy objectives. It reduces the number of motorway fatalities as experience in the US has proven. It saves energy and thus reduces CO_2 and other greenhouse gas emissions. It also reduces conventional pollution emissions such as HC and NO_x. Under most conditions, it also reduces road-tyre interface noise.
- Efforts to enforce existing speed limits are being increased. A more effective policy would be to either lower UK speed limits, require the installation of speed limiters on passenger cars, or both. Speed limiters would require EC legislation and are unlikely to be considered due to likely consumer unacceptability. Another method recently discussed by the Department of Transportation is to use new electronic technology to monitor the speed of passenger cars and collect a tariff for speeds over the existing speed limits. Public

response to the tax proposal has been critical due to its regressive nature and its apparent sanctioning of speeding if the operator wishes to pay for it.

8 Motor Vehicle Fuel Quality Regulations

One of the most successful air pollution reduction programmes implemented in any country is the reduction and removal of lead from petrol. The resulting rapid decrease in ambient lead levels and the relatively minor disruption in the petroleum refinery sector, despite dire warnings to the opposite, point to the regulation of transport fuel quality as a prime area for future environmental public policy.

Until relatively recently, fuel content was determined solely on its vehicle performance requirements. Recently, however, the experience with lead content restrictions is being extended to tackle other air pollution problems. Reduction of air toxics, sulphur, carbon monoxide, particulate matter, hydrocarbons, nitrogen oxides and corresponding decreases in smog are all being investigated via a fuel quality route. In addition, the greenhouse gas contribution of various fuels is the most recent environmental aspect of motor vehicle fuels receiving serious analysis. Most recently, high levels of ambient carbon monoxide have been reduced in the US through the addition of a minimal amount of oxygenate to petrol. The result has been nothing short of spectacular.

The following review outlines recent and future developments in fuel quality regulations around the world and the implications of those developments for the UK. As elsewhere in the report, this section outlines the potential future regulatory development concerning fuel quality. It does not attempt to assess the environmental necessity of making such changes nor the effectiveness of these measures relative to other potential policies.

8.1 Lead Content in Petrol

The content of lead in petrol (premium and regular) is regulated in many countries, including the EC. The share of unleaded petrol in the market is increasing as Table 31 shows below. Summing up the status of lead content in petrol, the following observations can be made:

Table 31 Market Share of Unleaded Petrol

Country By Region	1991 Percentage Share of the Gasoline Market			
	Total Unleaded	Super Plus	Super	Regular
Europe				
Austria	58.7	–	26.0	32.7
Belgium	37.4	7.6	29.8	–
Denmark	64.8	18.6	36.9	9.3
Eire	25.3	14.8	10.5	–
Finland	57.7	0.5	57.2	–
France	25.5	22.9	2.5	–
Germany	78.1	7.9	30.6	39.6
Greece	9.1	–	9.1	–
Iceland	67.0	–	15.9	51.5
Italy	7.0	–	7.0	–
Luxembourg	43.0	6.2	36.8	–
Netherlands	66.0	16.0	50.0	–
Norway	47.0	7.1	39.9	–
Portugal	8.4	–	8.4	–
Spain	3.3	0	3.3	–
Sweden	57.1	1.0	56.1	–
Switzerland	57.4	0	57.4	–
Turkey	0.6	–	0.6	–
UK[1]	46.9	4.9	36.0	
Americas				
Argentina[2]	–	–	–	–
Bermuda	74.2	74.2	–	–
Brazil	100.0	–	–	100
Canada[3]	100.0	–	–	–
Guatemala[4]	100.0	–	37.2	62.8
Jamaica[5]	15–20	–	–	–
Puerto Rico[6]	100.0	–	–	–
US[7]	96.5	–	–	–
Pacific Region				
Australia	35.7	–	0.6	35.1
Hong Kong	38.4	11.4	27.0	–
Japan	100.0	15.7	–	84.3
Malaysia	4.1	4.1	–	–
New Zealand	31.6	–	–	31.6
Singapore	43.8	33.6	–	10.2
Thailand	7.0	1.4	5.7	–

Notes: All figures rounded to nearest 0.1%. Some data are based on industry estimates and should be treated accordingly. Grade structure (Super plus, Super and Regular) approximates to European octane quality.

1 DoE, 1992, "Digest of Environmental Protection & Water Statistics," HMSO, London.
2 Not marketed in 1991.
3 Grade split unavailable.
4 Regular Grade 87 RON.
5 Estimate only – Industry figure unavailable.
6 Grade detail unavailable.
7 Grade split unavailable.

Source: Motor vehicle emission regulations and fuel specifications Concawe Report no. 2/92, Brussels, 1992

Status of Petrol Lead Content Regulations Around the World

- The UK introduced duty fuel tax differentials between leaded and unleaded petrol in 1987. Unleaded fuel in the UK now accounts for about 50% of petrol sold.
- In 1986, the US lowered the lead content limit in leaded petrol to 0.026 grams Pb/litre. This compares to the 0.15 grams Pb/litre content in the majority of EC countries (with the exception of Portugal and Greece which remain at the allowable 0.4 g/litre limit).
- In January 1992, California banned the sale of all leaded gasoline – premium and regular.
- All gasoline sold in Japan, Brazil and Canada is unleaded (1991). 97% sold in US in 1991 was unleaded.
- Leaded gasoline will be banned in the US from 1995.
- In Europe, Austria was the first to ban the sale of leaded gasoline – premium and regular (as of February 1993).
- Leaded regular is prohibited in several European countries (Belgium, Luxembourg, Denmark, Germany, Finland, Norway, Sweden, Switzerland) and in Australia and New Zealand.

Implications and Future Directions for the UK

- Regular grade leaded petrol will eventually be decreased in market share to the point where a ban for road transport use (excluding agricultural machinery) will become publicly acceptable. The increased use of auto catalytic converters will be the main driving factor.
- It is doubtful that allowable lead limits in leaded petrol will be reduced from the current 0.15–0.40 g/litre. Rather, the move of countries to prohibit leaded sales altogether will make such a move unnecessary.

8.2 Benzene Limits in Petrol

Benzene levels in petrol have increased recently as a substitute for octane-enhancing lead. Because of its carcinogenic nature and because mobile source exhaust is the primary contributor to benzene atmospheric concentrations, the benzene content in petrol is becoming increasingly limited.

The UK VOC Emission Inventory of 1991[95] indicates that benzene represented 2.51% of the total speciated VOC inventory at approximately 39,000 tons emitted. Some 95% of these emissions were attributed to petrol and diesel exhaust and petrol evaporation.

---
95 op cit note 9.

Limits of the benzene content in gasoline are relatively recent as is indicated in Table 32.

Table 32 Benzene in Petrol Content Limits

Country	Benzene Content % of Volume	Date Enacted
EC	5% maximum by volume for unleaded gasoline	unleaded from 1985, leaded grades from 1989
Austria	3% by volume	Effective Sept. 1990
Germany	Proposed 1% limit	EC denied tighter than EC standard
Italy	3.8% for unleaded	By agreement between oil companies and industry
	2.5% for selected summer-time markets for unleaded	From Feb–June '92 for selected Italian cities
	3.0% permanent standard	Proposed 12/91; to be effective from 1/93.
	2.0%	Proposed to be effective from 1/95. Pending EC approval
USA	1% volume maximum in reformulated gasoline market (see reform section)	From 1995–1996

Benzene levels in UK premium grade leaded gasoline, according to Associated Octel surveys, indicates that the UK fleet is operating on leaded fuel with, on average, less than the 5% EC content maximum.

From 1980 to 1990, the benzene content as % of volume ranged from 3.1 % in 1986 to 2.2% in 1987. Aromatics as % of volume ranged from 35.1% in 1980 to 30.7% in 1983 for the same decade. Thus, unlike other countries and fuel markets which have increased the level of benzene in fuels dramatically to enhance the octane level of unleaded petrol, this trend has largely not occurred in the UK.

Implications and Future Directions for the UK

- As the health effects of mobile source air toxics become a bigger part of the transport environment agenda in the EC and the UK, benzene content in gasoline will be a primary target for further regulation.
- It is unlikely that ambient benzene standards will be set, but guide values similar to those used for ozone and lead could be enacted at the EC level. It is more likely the fuel content restrictions will be

tightened. For the moment, the 1% maximum content by weight should be considered the reasonable near-term regulatory target.
- The implications on fuel volatility and other fuel quality parameters, as well as the implication on refinery production of a lower benzene content, should be investigated.

8.3 Reformulated 'Cleaner' Gasoline

During the Clean Air Act debates in the US, serious consideration was given to mandating a minimum number of alternatively-fuelled automobiles and the fuel infrastructure for these alternative-fuelled vehicles to operate on. For the US refinery industry which did not control many of the alternative fuel feedstocks (natural gas or agricultural stock), this was a potential threat. Faced with this development, the industry, lead by ARCO, promoted the idea that a 'cleaner' form of gasoline could be produced and that non-crude alternative fuels would not have to be mandated. They claimed that the new "reformulated" gasoline would accomplish much of the same benefit of alternative fuels such as methanol and ethanol – namely lower HC, NO_x and CO emissions.

The debate that followed and the subsequent 'regulatory negotiation' between government, auto manufacturers, refiners and environmental groups provided a fascinating chapter in US environmental history. The main outcome is that the US re-shaped the petrol market in a significant way.

As specified in the Clean Air Act and subsequent negotiations, US reformulated gasoline is being introduced in several areas of the country suffering from the worst tropospheric ozone. 'Reform', as it is called, must meet the content and performance specifications outlined in Table 33 below.

In addition to these specifications, the reformulated gasoline must have certain performance specifications.
- *NO_x emissions*: no increase; can be met by limiting oxygen content to less than 2.1% or using MTBE of not more than 2.7% mass oxygen.
- *Air Toxics reduction* of at least 16.5% (defined as benzene, 1-4, butadiene, polycyclic organic matter, formaldehyde and acetaldehyde).
- *VOC reduction* of at least 16.5% from 1997 (including both evaporative and exhaust emissions). Fixed RVP limits from 1995–1996.

Table 33 US Reformulated Gasoline Specifications

Specification Item	Target	Refinery Average Limit	Absolute Limit
RVP Class A*	7.2 psi/49.7 kPa	7.1 psi/49.0 kPa	7.4 psi/51.1 kPa (max)
Class B*	8.1 psi/55.9 kPa	8.0 psi/55.2 kPa	8.3 psi/57.3 kPa (max)
Oxygen	2.0% mass	1.2% mass	1.5% mass (min)
Benzene	1.0% vol	0.95% vol	1.3% vol (max)
Heavy Metals	None without waiver from the EPA		
T90E	Average no greater than refiner's 1990 average		
Sulphur	As above		
Olefins	As above		
Detergent additives	Compulsory**		

* 16.5% relative to 1990 average gasoline
** Federal requirement of Clean Air Act but not yet defined by the EPA

Source: Concawe Report no. 2/92, 1992

In addition to the Federal US standards, California also adopted its own set of gasoline specifications. In effect from January 1992, Phase I standards include: summer RVP limit of 7.8 psi (53.8 kPa) maximum; leaded gasoline ban; and deposit control additives.

California Phase II gasoline standards, to be in effect from 1997, are as follows:

Table 34 California 'Phase II' Reformulated Gasoline Specifications

Parameter (Maximum Limits)	Producer Limit[1]	Averaging Limit	All Gasolines Limit
Volatility (psi)[2]	7.0 psi/48.3 kPa	–	7.0 psi/48.3 kPa
Sulphur (ppm)	40	30	80
Aromatics (vol%)	25	22	30
Benzene (vol%)	1.0	0.8	1.2
Olefins (vol%)	6.0	4.0	10.0
Oxygen (wt%)[3]	1.8–2.2	–	2.7 max / 1.8 min
T90 °F (°C)	300 (149)	290 (143)	330 (166)
T50 °F (°C)	210 (99)	200 (99)	220 (104)

1 Limits set on each batch of fuel produced if averaging is not used.
2 Summertime only, variable depending upon location.
3 Final decision on this requirement deferred until after the date publication of this report.

Source: California Air Resources Board as reported in CONCAWE Report no. 2/92, 1992

Summary Points from the US Reformulated Gas Experience

- Reformulated gasoline represents a new element in road transport emission control policy – a switch from concentrating solely on the car and lorry itself, to the fuels that are used by vehicles.
- This development came about because of increased interest in the US Congress to mandate the use of 'alternative' fuels (methanol, ethanol, CNG, others) on a limited initial basis.
- The reformulated gasoline development was initially an industry-led approach designed to maintain market share.
- From an air quality perspective, significant reductions in ambient air pollution problems are expected. At this early stage, however, there is disagreement about the extent of potential reductions.
- Although reformulated gasoline's initial purpose was to help reduce tropospheric ozone levels, major benefits in the control of hazardous air pollutants will be gained.
- Early evaluation of the Federal programme raises several questions. Mixing of ethanol based fuels with MTBE-based fuels by consumers could result in higher VOC emissions than that which would result from either fuel alone. In addition, concerns exist about the use of oxygenates at levels greater than 2% (which is encouraged by the present approach). NO_x emissions and toxics are of concern.
- Another concern of using MTBE is potential elevated MTBE blood levels humans in areas where the fuel additive is used.
- The obvious advantage of addressing air pollution from the fuel perspective (in addition to vehicle-control itself) is the ability to immediately reduce pollution from the entire fleet rather than from new vehicles alone.

Implications and Future Directions for the UK

- The ability of 'cleaner' petrol to assist in the reduction of NO_x emissions should be closely monitored. If this aspect is successful, then UK and EC fuel reformulation could develop along these lines. Recently, DG XI of the EC began discussions to encourage the introduction of reformulated fuels voluntarily and through the possible use of fiscal incentives. Further development, to coincide with the next stage of vehicle requirements, is also being discussed.
- As knowledge of the ambient and urban levels of hazardous air pollutants become increasingly known in the UK, and as the health effects of these pollutants become more of a concern for UK policy

makers, the attractiveness of reformulating fuels from an air toxic perspective is enhanced.

- Volatility control for environmental reasons will take on greater prominence in the EC and will eventually result in significantly reduced petrol volatility. Quantification and regulation of evaporative emissions from vehicles will also act to create pressure from the auto manufacturers for lower volatility fuels.
- The present use of averaging fuel content specifications, rather than requiring every gallon of fuel to be made to the same specification, significantly reduces the potential cost of fuel reformulation and is a major advantage in terms of managing supply and demand of various fuels to various markets. Averaging proved very useful in introducing unleaded gasoline in the US market, lowering expected industry costs from $4 billion to less than $400 million, resulting in none of the reported fuel shortages that were predicted by the US refinery industry and producing one of the most cost-effective ambient pollution control policies ever.

8.4 Diesel Fuel Quality

The Increased Demand for Diesel Fuel

UK road transport consumed some 10.65 million tonnes of diesel in 1990, up from 5.85 in 1980. This amounted to approximately 30% of total petroleum consumption of all road transport in the UK compared with 23% in 1980[96]. In the OECD as a whole, diesel demand was approximately 40% of the market in 1988 and is projected to rise to 45% motor fuel demand by the year 2005. This is approximately a 50% rise in absolute demand from 80 million tons in 1988 to 121 million tons in the year 2005[97].

The growth in the diesel fuel market is driven by the estimated increase in the diesel passenger vehicle market. One future-looking study utilized the following market size for the diesel in the UK (see Table 35).

Implications for Refineries

This increased demand for diesel fuel in the OECD and the UK will increase pressure on refiners to expand the middle range of the crude oil distillation column. This will reduce the amount of crude oil currently devoted to production of lubricating oils and asphalt. The pressure to create a more environmentally benign diesel fuel, with lower sulphur content, lower particu-

---

96 The Department of Transport, 1991, *Transport Statistics Great Britain 1991*, HMSO, London.

97 In 1989, Belgium's diesel market was 53% of the total petroleum market, Spain and Italy 51%, France 45%, Denmark 43% and the Netherlands 42% according to *Transport Annual Statistics 1970–1989*, Eurostat, Brussels, 1991.

Table 35 Assumed penetration of diesel cars in the UK new car market

Year	% New cars diesel
1990	6
1995	12
2000	16
2005	20

Source: Eggleston, 1992

late emissions and lower PAH emissions, will increase the demand for low sulphur crude feedstocks. Due to the limited and declining availability of 'sweet' (low sulphur) crudes, there will be an increased demand for refinery hydrocracking process capacity. In addition, the pressure to use products of secondary conversion processes in the diesel blends (with the exception of the products of hydrocracking), will tend to result in a fuel with a higher aromatic content and lower cetane index number.

This in turn could result in increased government regulation of both the aromatic content (eg 10% limit in California 1993 diesel; 5% limit in Swedish Class 1 diesel) and minimum cetane indices (eg 46 in Denmark, 48 in California, 50 in Swedish Class 1) in order to control potential increases in PAH and particulates in exhaust. The incorporation of both a specified aromatic content and minimum cetane in diesel fuel quality regulations is deemed necessary by some. This is to avoid the potential problem of rising aromatic content due to the use of additives which are designed to boost falling cetane indices. The impact on refiners of increased specification, however, especially of aromatic content, will severely limit the pool of crude oil that can be incorporated into diesel fuel. This will increase the overall cost of diesel refining due to the necessity to employ more costly refinery hydrocracking capacity.

Diesel Fuel Quality Standards and Regulations

Diesel fuel quality regulations are becoming quite prominent in industrialized economies. The bulk of the fuel quality regulation has centred on the sulphur content of diesel measured as a percentage of mass. Recently, other fuel composition specifications, such as limits on aromatic content mentioned above, have become more common place.

Sulphur limits for fuel oil, including diesel, have now been set in the EC. They were originally agreed to in March 1992. Existing EC law allows a maximum sulphur content of 0.3% by weight. This will be reduced for diesel fuel to 0.2% by October 1, 1994 and to 0.05% by October 1996.

Table 36 below indicates current or planned sulphur content regulations around the world.

Table 36 Diesel Fuel Sulphur Content Limits

Country	Current Sulphur Content (% by weight)	Proposed or Future Content
UK	0.3%	As part of EC, 0.2% by Oct '94 0.05% max limit by Oct '96
EC	0.3%	0.02% by Oct '94 0.05% by Oct '96
Belgium	0.2%	Effective 10/89
Denmark	0.2%	Effective 11/88
Ireland	0.3%	
Italy	0.3%	0.2 under consideration
France	0.3%	
Luxembourg	0.2%	Effective 10/89
Netherlands	0.2%	Effective 10/88
Spain, Greece, Portugal	0.3%	
Austria	0.15%	Effective 1986
	0.05%	From 10/95
Brazil	0.7%	From 1988
Finland	0.2%	From 1989
India	0.5%	
Japan	0.2%	From October 1992
Sweden	0.001%, 0.005%, 0.3%	
Switzerland	0.05%	from January 1, 1994
US		0.5 current
	0.05%	mandatory in California now, all US by Oct '93
Canada	0.4%	Effective from 1989

Other Diesel Fuel Quality Specifications

In addition to sulphur content of diesel fuel, other characteristics are coming under scrutiny – some for environmental reasons. The diesel fuel regulations in California and Sweden are the most comprehensive set of specifications for such purposes.

Sweden

The Swedish fuel characteristics are described in Table 37 below. They were made effective in January 1992 and incorporate a diesel fuel quality classification system based on environmental criteria. The programme is being implemented with the aid of differential tax levies.

Table 37 Sweden 1992 Diesel Fuel Classifications

Fuel Characteristic	Urban Diesel 1	Urban Diesel 2	Standard Grades
Sulphur Content % mass (max)	0.001	0.005	0.30
Aromatics % vol (max)	5	20	–
PAH % vol (max)	0.02	0.1	–
Distillation:			
IBP (min) oC	180	180	–
10% (min) oC	–	–	180
95% (max) oC	285	295	
Density (kg/m3)	800–820	800–820	
Cetane Number	50	47	
Tax Rate ($/m3)	135	168	210
Tax Rebate per litre	0.052 ECU	0.029 ECU	0

Source: Concawe 1992, Sweden, 1993 (see note 98 below)

Denmark

Denmark has also introduced a diesel fuel improvement programme with tax incentives for buses as indicated in the following table.

Table 38 Diesel Fuel Introduction in Buses in Denmark

Characteristics	A CEN quality	B CEN 0.05% S quality	C Public Bus Service Ultra Light Diesel
Sulphur max % mass	0.20	0.05	0.05
95% Distillation (°C)	370	370	325
Density	820–860	820–860	820–855
Cetane Number	49	49	50
Cetane Index	46	46	47
Tax (DKr/l)	2.04	1.94	1.74
Tax Incentive	0	0.10	0.30

Source: Concawe Report no. 2/92, 1992

California

In the US and separately in California, diesel fuel quality specifications other than sulphur content have recently been introduced. The California specifications, in addition to aromatic content and cetane index specs, contain a specification for polycyclic aromatic hydrocarbons and nitrogen mass content. Unlike recent developments for gasoline, diesel fuel specifications are not formulated as performance specifications.

Table 39 California Diesel Reference Fuel Specification

Property	Limit
Sulphur % Mass (max)	0.05
Aromatics % volume (max)	10.0
Polycyclic aromatics, % mass (max)	1.4
Nitrogen, ppm mass (max)	10.0
Cetane number, min	48.0

Source: Concawe Report no. 2/92, 1992

Europe

Recent CEN diesel fuel specifications incorporate a minimum cetane index (46) and a maximum sulphur content (ultimate goal of 0.05% to align with EC directives), but not a limit on aromatic content.

Clean Diesel Tax Policies

A number of countries have incorporated financial incentives in the form of either reduced taxes for cleaner diesel or increased taxes for standard diesel fuel oil. These are summarized in Table 40.

Table 40 Clean Diesel Tax Incentive Policies

Country	Sulphur Content (% mass)	Aromatic Content Limit	Minimum Cetane Index	% lower fuel taxes compared to 'standard' diesel
Norway				
Class 1	< 0.05%			100%
Class 2	0.05–0.25%			50%
Denmark	0.05		47	15%
Sweden				
Class 1	0.001	5%	50	36%
Class 2	0.005	20%	47	20%
Switzerland	0.05			probable tax incentive when implemented in 1994

Comment

The reduced tax tariff for cleaner diesel fuel is intended both to increase consumption of the "cleaner" fuel by eliminating the production price premium and, in some cases, provide a price discount over 'baseline' standard fuels. Unfortunately, the amount of potential price discount and the price premium of diesel fuels was unavailable to include in this report. The impact of the Swedish programme on fuel demand has been evaluated, however. According to a recent summary of this taxing policy, the effects have been dramatic in Sweden[98]. The diesel fuel oil tax became effective in 1991 and is part of a wider set of environmental taxes covering carbon dioxide, sulphur and nitrogen oxides. Since 1991, diesel fuel oil has been divided into three classes –

---

98 Swedish Environmental Protection Agency, 1993, *Recent Events in Sweden*, Energy Section, to the OECD Air Management Policy Group.

one for urban consumption with improved environmental qualities and one for standard class. Fuels in the urban class are taxed less than the standard fuel in order to stimulate demand.

In 1990, less than 1% of the diesel oil sold would have qualified under classes 1 or 2. By the end of 1992, this amount had increased to 15% for class 1 and 60% for class 2. A new refinery facility to produce the top class diesel fuel has opened in Sweden with capacity to serve 25% of current Swedish total diesel demand.

Implications and Future Directions for the UK

- Based on developments elsewhere, a move to 0.05% maximum diesel fuel sulphur content is most likely.
- Diesel fuel specification is becoming increasingly specific with limits on aromatic content and cetane prevalent.
- As the content of diesel becomes more specified, averaging of fuel specifications across the amount of fuel produced and sold should be allowed to maintain efficiencies.
- As diesel fuel quality specifications become more specific, pressure will grow on governments to move to diesel fuel 'performance' standards, not unlike the reformulated gasoline program in the US, in order to maintain refinery flexibility and to ensure sufficient capacity for the growing diesel market.
- The RCEP, in its report on diesel emissions, recommended that the UK immediately adapt a diesel fuel price differential tax that would more than offset the increased 0.05 sulphur content production price premium in order to increase the market share for lower sulphur diesel fuel in the UK[99]. The government rejected such a policy in its response to the RCEP. This approach should be given serious reconsideration.

8.5 Oxygenated Fuels

Many countries have allowed or mandated the use of various oxygenate additives to gasoline. This has been done to help control hydrocarbon and carbon monoxide emissions and to supplement crude oil supplies in the road transport fuel market. To ensure standardized fuel quality, many countries (including the EC member states) have regulated that amount of oxygenate additive that can be added to petrol. Some countries have mandated certain minimum oxygenate levels (Brazil and US) for purposes of emission control or domestic energy policy.

99 Royal Commission on Environmental Pollution, 1991, *Emissions from Heavy Duty Diesel Vehicles*, 15th Report, London.

EC

Table 41 below outlines the oxygenate regulations from EC Directive 85/536/EEC.

Table 41 EC Oxygenates limits (85/536/EEC)

	A (% Vol)	B (% Vol)
Methanol, suitable stabilizing agents must be added (a)	3%	3%
Ethanol, stabilizing agents may be necessary (a)	5%	5%
Iso-propyl alcohol	5%	10%
TBA	7%	7%
Iso-butyl alcohol	7%	10%
Ethers containing 5 or more carbon atoms per molecule (a)	10%	15%
Other organic oxygenates defined in Annex Section 1	7%	10%
Mixture of any organic oxygenates defined in Annex Section 1 (b)	2.5% weight, oxygen, not exceeding the individual limits fixed above for each component	3.7% weight oxygen, not exceeding the individual limits fixed above for each component

Notes:
(a) In accordance with national specifications or, where these do not exist, industry specifications.
(b) Acetone is authorized up to 0.8% by volume when it is present as a byproduct of manufacture of certain organic oxygenate compounds.

Proposed New Limits
In its June 1992 working document, the Commission has proposed the following changes:

Column A: Ethanol: From 5 to 10% volume
           Ethers: From 10 to 15% volume
           Oxygen: From 2.5 to 3.5% Weight
Column B: Ethanol: From 5 to 10% volume

Source: Concawe Report no. 2/92, 1992

The reader should note the important changes currently being reviewed in the form of a Commission working document. These regulations are intended to remove barriers to the addition of oxygenate additives. Levels in column A indicate maximum levels of additives that must be allowed (not mandated). Column B indicates levels that would also require clear notification at point of purchase of the presence of the additive.

This regulation was adopted by the UK in 1989. Market shares of oxygenated fuels in the UK were not available.

US

The US has ruled that certain oxygenate additives may be added to petrol in amounts that do not exceed 2.7% mass. Waivers to this percentage rule have been granted to a number of fuel additives, including but not limited to:

- 'Gasohol' consisting of gas with 10% by volume ethanol which contains 3.5% oxygen. This was instituted in 1979 to assist US grain producers who were experiencing market shortfalls.
- Methanol up to 5% by volume plus co-solvent additive.
- MTBE up to 15% by volume which has facilitated the use of MTBE for the new US reformulated gasoline market.
- In addition, to control CO ambient levels, some 41 cities representing 31% of total US gasoline sales volume are required to add oxygenates so all gas has a 2.7% mass oxygen content.

Brazil

Brazil, for purposes of national energy policy, implemented the world's most far-reaching petrol-substitution programme in 1979. At that time two grades of fuel were made available to the public. The Brazilian car market responded by supplying vehicles capable of running on the new fuels. The 'E93' fuel was a neat hydrated ethanol fuel containing 93% ethanol whose feedstock is primarily sugar cane and the second grand 'gasolina' is a blend containing 22% ethanol. Market penetration and shares were not available.

Increased Formaldehyde Emissions of Oxygenated Fuels

A subject of concern with the implementation of oxygenated fuels and fuel additives is the potential for increased formaldehyde emissions and potential adverse health effects associated with formaldehyde. As part of the new US oxygenated fuels, reformulated gasoline and alternative fuelled vehicles programmes, the EPA has made estimates of increased emissions of formaldehyde[100]. The results, listed in Table 42, indicate that formaldehyde emissions of oxygenated fuels were greater than for non-oxygenated fuels.

---

100 R Cook, 1992, *Calculation of exhaust and evaporative toxic emissions mass fractions for the motor vehicle related air toxics report*, US Environmental Protection Agency, Office of Mobile Sources, Ann Arbor.

Table 42 15% MTBE and 10% Ethanol Emission Fraction Adjustment Factors for Formaldehyde

Vehicle Class	Catalyst Technology	15% MTBE Adjustment Factor	10% Ethanol Adjustment Factor
LDGV/LDGT	3-way	1.6746	1.4758
LDGV/LDGT	3-way + ox	1.2672	1.2288
LDGV/LDGT	oxidation	2.0244	1.2400
LDGV/LDGT	non-catalyst	1.5256	1.1034

LDGV = light duty gas vehicle
LDGT = light duty gas truck

Implications and Future Directions for the UK

- The use of increased amounts of oxygenate in motor vehicle fuels in the UK will most likely mirror the development of oxygenated fuels policies in the EC.
- If exceedence of carbon monoxide ambient standards and guidelines continues or increases, it can be expected that mandated levels of oxygenates in petrol will be seriously looked at as a regulatory solution.

8.6 Alternative Fuels

US

As mentioned above, when the Clean Air Act was being debated in Congress, original versions drafted by the Administration contained proposals for a minimum number of 'alternative-fuelled' vehicles to be sold around the country in the worst polluted areas. The mandated availability of fuels for these vehicles was also proposed. Much of the early 'alt-fuels' debate was over the merits of methanol and ethanol. In the end, the US Congress opted for a reformulated gasoline programme and moved away from a mandated alternative fuelled vehicles programme. In its place, however, a California alternative fuels pilot programme and an US alternative-fuels fleet programme were implemented and will take effect in 1996 and 1998 respectively.

In addition, the new Act defines performance specifications for what qualifies as 'clean alternative fuels.' These specifications will apply to the Clean Fuel Fleet programme, whereby centrally-fuelled vehicle fleets of ten vehicles or more in size, located in the worst polluted areas, will have to have emission performance characteristics specified below by the year 1998. The performance specifications mandated for the clean fuels '*fleet*' programme are the same as the California low emission vehicle (LEV) programme:

NO$_x$	0.2 g/mile
NMHC	0.075 g/mile
CO	3.4 g/mile
Formaldehyde	0.5 g/mile

The California clean fuel *vehicles* programme will begin in 1996 when 150,000 clean vehicles must be made available for sale. This number will increase to 300,000 in 1998. Like the federal fleet programme, alternative fuel vehicle requirements can be met by either producing vehicles designed to run solely on non-petrol fuels (CNG, LPG, electric, methanol, ethanol, hydrogen, etc) or vehicles that, together with alternative fuels, can demonstrate certain performance specifications. The performance specifications for the California programme are the same as the California Transitional Low Emission Vehicle Programme.

NO$_x$	0.4 g/mile
NMHC	0.125 g/mile
CO	3.4 g/mile
Formaldehyde	0.015 g/mile

The effect of using alternative fuels on air emissions has been evaluated in a number of studies. A look at the effects of methanol on changes in air toxic emissions is presented in Table 43. It shows decreased levels in all toxics from using either 100% methanol or an 85% methanol blend, with the exception of primary formaldehyde.

Table 43 Percent Change in Air Toxics Levels for M85 and M100 Relative to Gasoline

Pollutant/Source	Auto Oil Study[101]	EPA Methanol Special Report[102]	
	M85	M85	M100
Exhaust Benzene	−84	−77	−99
Evaporative Benzene	–	−67	−100
Running Loss Benzene	–	−69	−100
Refuelling Benzene	–	−14	−100
Other Gas Refuelling Vapors	–	−14	−100
Exhaust 1,3-Butadiene	−93	−64	−99
Exhaust Gasoline POM	–	−72	−99
Primary Formaldehyde	+436	+600	+200
Secondary Formaldehyde	–	−43	−80

Ethanol also exhibits improved air toxic emission characteristics with lower benzene, 1,3-butadiene, refuelling vapours and particulate[103].

8.7 Ethanol and Biofuels: Agricultural Policy, Energy Security and Global Warming

The catalyst behind the current alternative fuelled vehicle programmes (and for increasing the level of oxygenates in petrol) is for the reduction of emissions of HC, NO_x, CO and corresponding decreases in ambient pollution problems.

A recent initiative to expand the use of biofuels or ethanol in the EC is also driven by these same concerns, but is primarily related to agriculture and energy policy goals – the same catalysts for the original 'gasohol' programme in the US and the Brazilian ethanol programme.

The EC in March of 1992 issued a draft directive to reduce the amount of Member State tax on biofuels (ethanol, methanol) to a maximum of 10% of current petrol and diesel tax levies. The goal is to create a 5% market share for biofuels by the year 2005 – involving the use of seven million hectares of land to produce eleven million tonnes of biofuel.

---

101 Auto Oil Study Research Programme, 1992, *Emissions and Air Quality Modeling Results from Methanol/Gasoline Blends in Prototype Flexible/Variable fuel Vehicles*; Technical bulletin No. 7.
102 US Environmental Protection Agency Special Report, 1989, *Analysis of the Economic and Environmental Effects of Methanol as an Automotive Fuel*.
103 US Environmental Protection Agency Special Report, 1990, *Analysis of the Economic and Environmental Effects of Ethanol as an Automotive Fuel*.

The EC proposal is being evaluated in terms of its global warming contribution. This is unlike the Brazilian and US ethanol programme and the new reformulated gasoline programme in the States which are evaluated primarily along the lines of energy security or agricultural policy. A recent study produced by ERL for a lobby of fuel oxygenate producers analyzed the CO_2 factor of ethanol[104]. This life-cycle analysis, accounting for the energy used in the various ethanol production processes, concluded that there is a minor CO_2 emissions penalty in most ethanol production cases. When modelled on a European-wide scale (one million tonnes substitution of bioethanol for petrol), there would be a miniscule effect on EC CO_2 levels (a net reduction of 0.002%).

The results of any such analysis must be considered in the appropriate context and eventually expanded to incorporate other environmental effects. Basing a decision to switch to ethanol over petrol for the purpose of CO_2 reduction alone is as short-sighted of a decision as switching fuels on the basis of smog-reduction potential alone. Because fuel production and use contributes to more than one air pollution problem – not to mention the water and waste pollution created by exploration, production, refining and fuel distribution – policy makers must be careful not to attempt to utilize the environmental banner for the advancement of other forms of policy, like agriculture and energy security.

It is not difficult to develop the appropriate scope for a life-cycle analysis that incorporates all of the environmental goals that a government is trying to achieve, but it is quite difficult to assemble and carry out that analysis. Conducting such an analysis while attempting to consider other policy areas, such as energy, transportation, security, and agricultural policy, is even more daunting.

When considering some of these other factors as the ENDS report did, the case for ethanol becomes even cloudier:

- Ethanol production in Brazil and the US, the two largest scale programmes in the world, have achieved an average price of approximately $40/barrel compared to average petrol prices of $20/barrel. Large production subsidies and tax incentives have been used to support the US ethanol programme – kept in place by a powerful mid-States agricultural grain producing lobby.

104 Environmental Resources Ltd for European Oxygenated Fuels Association, September 1990, "Study of the environmental impacts of large scale bioethanol production in Europe" as reported in *Brussels and the bioethanol boondoggle*, pp 22–25, ENDS Report 212, September 1992.

- Without major CAP changes, large scale European production would incur an additional per/acre 'set aside' costs because the crops would be pulled out of food production.
- Without fuel tax changes, the incentive for increased demand for biofuel would carry with it a tax revenue loss factor – estimated at about £2.8 billion/year across the EC when the programme reaches a 5% market share.
- Justified on a CO_2 reduction basis alone, there may be other more cost-effective means to reduce CO_2, such as reforestation programmes.

Implications and Future Directions for the UK

- Any increase in the use of biofuels as a substitute for petrol and diesel in the EC and UK will be part of a complicated policy development involving environmental, energy, and agricultural objectives.
- The evaluation of such a programme in the EC will take place on broad terms, utilizing life-cycle analysis. Under such scrutiny, the environmental benefits from a global warming perspective may not be immediately obvious.
- If, however, the programme is evaluated from an even broader environmental perspective, the merits of its implementation may be increased.
- It is likely that agricultural policy, not environmental policy, will eventually be more influential in determining the future of a wide-scale EC biofuels initiative.

9 Conclusions

This report has described how road transport contributes to one of the most pervasive environmental problems on today's agenda – air pollution. By outlining the changing nature of road transport air pollution in the UK, the report illustrates a future agenda and the likely policy and technical responses.

Policies aimed at motor vehicle pollution are unique. Unlike most other types of environmental policies, those developed to address vehicle emissions have the unique ability to replicate themselves world-wide – albeit at different times and in different forms according to the geographic markets considered. To improve the UK's preparation, it is vital to look at the past trends, current positions and future directions of vehicle emission control policies around the world.

By noting the direction that transport environmental policy and technology may develop and by improving the understanding of air pollution within the UK, the future challenge can be outlined and the strategic capacity to respond enhanced. Without an integrated strategic assessment of future transport policy, opportunities for an improved quality of life and economy will simply not be realized. This concluding chapter highlights the main summary points from the body of the report. Key implications for the UK government and industry are highlighted.

9.1 Foresight and Transport

A number of observations developed during the first phase of the Environmental Foresight Project (see Volume 1) are worth mentioning in consideration of the UK's future transport-related environmental agenda.

- Transport environmental problems, like all modern and future environmental problems, will need an integrated solution.
- The ultimate solution to environmental problems, including transport, lies in an understanding and management of the economic and social conditions that are the driving force behind such problems.
- Energy and environmental policy will be even more closely linked in the future. Transport environmental policy is a perfect example of the necessity to link energy and environmental concerns.
- As identified in the Foresight Project, the scope of environmental policy is becoming both smaller and larger. In the case of transport,

global pollution controls are now active and in the future, urban 'micro' environmental conditions will become the subject of concern.
- The UK's future transport and environment agenda is for the most part known. This confirms the project's general observation that the prediction of new environmental problems is difficult and not always the most useful type of foresight. Preparation, rather than prediction, can provide more tangible results.
- Human health concern is a major force for change for new environmental issues in the UK. Knowledge of the health effects of mobile source air pollutants and evidence of these pollutant's concentration in the atmosphere and in urban conditions is increasing. This will lead to a renewed emphasis to combine public health policy with environmental health policy in the UK.
- Like other environmental problems, road transport air pollution policy is going through a natural progression from the most obvious source of the problem to less obvious and smaller sources. Just as non-point water pollution sources will be brought under control in the future, smaller sources of road transport air pollution will come under control.

9.2 Road Transport and Air Quality

Concerning the contribution of motor vehicles to air pollution in the UK, the following observations can be made:

- Motor vehicles are a major contributor of a number of air pollutants, including noise, in the UK in 1990:

Carbon monoxide (CO)	90%
Nitrogen oxides (NO$_x$)	51%
Black smoke (BS)	46%
Volatile organics (VOCs)	41%
Carbon dioxide (CO$_2$)	38%

- Motor vehicle air pollution will be reduced dramatically through the adaptation of 3-way catalysts on passenger cars and other emission regulations. Despite this progress, motor vehicles in the UK will continue to be a major source of air pollution in the year 2000 and some pollutant emissions will be on the rise again after the year 2000 due to a growth in vehicle miles travelled. In the year 2000 road transport will contribute the following percentage to UK air pollution.

Nitrogen oxides	46%
Volatile organic chemicals	33%
Carbon dioxide	20%

- The contribution of heavy goods vehicles (HGV) to UK road transport air pollution emissions will increase over time as passenger cars become cleaner:

Nitrogen oxides	Emissions will increase from 35% of road transport emissions in 1990 to 48% in 2010.
Volatile organics	Emissions will increase from 14% in 1990 to 17% in 2010.
Particulate matter	Emissions will remain constant at 66% of total road transport emissions over time.

9.3 Road Transport Policy Trends

The amount of air pollution generated by road transport in the UK is directly related to emission control policies of all types. The future UK policies addressing motor vehicle pollution will be strongly influenced by policy developments elsewhere. These policies are part of clearly discernible trends which can be projected into the future.

Concerning air pollution from passenger cars, some of the major trends include:

A Changing Policy Goal
Emission control policies originally aimed at conventional air pollutants (eg NO_x, HC, CO) have expanded to address global pollutants (eg CO_2, CFCs) and hazardous air pollutants (eg benzene, 1,3-butadiene).

A Changing Scope of Motor Vehicle Policy
Motor vehicle air pollution was originally considered a local problem. Concern has expanded to a regional context with the improvement of air chemistry knowledge and is now global. The next scope for policy development is likely to be at a local 'micro' environmental level. This is due to the increased concern over elevated levels of air pollutants found in certain urban situations such as near roads and within vehicles themselves.

A Move towards Local Air Quality Management
Due to inadequate regional or national management of air quality, local air quality management areas have developed with planning responsibilities and the capability to implement local control measures.

Expanding Focus of Emission Controls

Vehicle emission control programmes have expanded from the tailpipe to other sources of vehicle pollution not addressed previously. 'In-use' emissions are being regulated through vehicle inspection and maintenance programs and 'evaporative' emissions are being controlled through modified vehicle testing and certification procedures.

From Vehicles to Fuels to Fuel Distribution

In addition to the expanding control of emissions emanating from the vehicles themselves, vehicle fuels are becoming controlled. Both fuel quality and mandates for alternative non-crude transport fuels are being considered and implemented.

From California to Europe

Motor vehicle emission control policies for the past 25 years have been initiated in California first, next adopted by the US as a whole, and then by the EC, Japan and Scandinavia. This trend should continue for the next decade. Traffic planning measures, however, are more likely to originate in Europe first. Europe also has a stronger lead in integrating energy and environmental policy which, in the case of transport, leads to developments such as the life-cycle energy analysis of vehicle fuels, and the progressive use of fuel taxes to limit demand.

From Cars to Lorries to Fleets

Passenger cars have received the most pollution control policy attention. As they become relatively more controlled, attention will switch to less controlled sources which make up a relatively greater proportion of road transport emissions, such as lorries. As alternative fuels become a subject for policy consideration and as air quality management switches to local jurisdictions, policies addressing fleets of vehicles will become commonplace.

From Vehicle Performance to Vehicle Use

As the technical options for vehicle emission control are implemented, attention will turn to the emission performance of vehicles under different driving conditions. Vehicle speed will be most likely affected.

From Vehicle Regulation to Transport Planning

Eventually, the growth in either the number of vehicles or the miles that they drive will result in the necessary application of transport planning and various restrictions on either the demand for private transport or the supply of it.

From Mobile to Stationary Sources

As mobile sources are progressively more controlled, and as air pollution problems traditionally associated with mobile sources persist, stationary sources of air pollution will come under control. This integration of mobile and stationary source control measures will allow for a more efficient set of control policies.

By noting and evaluating these and other trends, the UK will be in a clear position to evaluate the necessity for such new policy developments, given its own pollution problems. Some of these developments, necessary or not, will affect the UK market and policy arena. By understanding the direction of world transport environmental policy, the UK will be able to prepare for the implementation and technical requirements that these trends represent.

9.4 Potential UK Policy Developments

Considering specific details of the general trends noted above, the following policy developments have the potential to occur in the UK and the EC in the future. This assessment of potential development is based on the assumption that policies developed elsewhere are very likely to be more or less universally adapted. If not adapted, they will be experimented with. The following assessment, that the policies could be implemented in the UK, is made independent of an assessment of their necessity for a particular UK air pollution condition. This is not to say that they will not be necessary for pollution control or advantageous to economic competition. In most cases they will be.

This conclusion reviews developments in the following areas:

- Future public health concerns
- The new global agenda
- Future tailpipe emission control policy
- Vehicle emission controls beyond the 'tailpipe'
- Changes in vehicle fuel policy
- New policy responses

9.4.1 Future Public Health Concerns

Motor Vehicle Emission Health Effects

Conventional Pollutants

- Most of the epidemiological research (either occupational or community-based) has centred on acute health effects of short-term exposure to relatively high levels of pollutants such as NO_2, SO_2, and O_3.
- As new research defines the health effects of long-term, lower level exposures, it can be expected that current ambient air pollution standards will be seriously reviewed for the adequacy of protection they offer.
- Health effects of particular concern include low levels of particulate matter-related acid aerosols and increased morbidity, and low levels of O_3 and NO_2 and increased respiratory health effects among sensitive populations (asthmatics, exercising adults) and general populations.

Hazardous Air Pollutants

- Mobile source air emissions increase the incidence of cancer and non-cancer health effects in the general population. This effect is increased for urban populations. Recent studies quantifying the relationship between levels of air toxic mobile emissions and cancer have been completed in a number of countries.
- The per-vehicle cancer incidence associated with the UK fleet could be greater than that associated with studies of mobile source air pollution in other countries (eg US). This is due to the low number of catalyst controlled vehicles and the relatively higher percentage of diesel-fuelled cars in the UK fleet.
- It is likely that a mobile source air toxics control programme will be implemented in the EC and the UK in the near future. It is likely that the programme will initially concentrate on the contribution of motor vehicle fuel to air toxic emissions. Benzene and 1,3-butadiene could be chemicals of concern.

Local 'micro' environmental concerns and controls

- As kerb-side and in-vehicle pollutant level monitoring increases and risk estimations are made, it is likely that specific regulations will be developed for these urban 'micro' environments. The programmes could contain occupational exposure standards where applicable (eg in garages, toll booths, service stations), health advisories, and local traffic planning.

9.4.2 The New Global Agenda

Vehicle Fuel Efficiency

- The concern for global warming, combined with existing reasons for improved vehicle fuel efficiency (agricultural policy, energy security policy, oil-price shock), will result in renewed policy efforts to improve vehicle efficiency.
- Further increases in fuel efficiency standards in the US and Japan will create a demand for more fuel efficient vehicles by all manufacturers and could create potential opportunities for increased foreign sales of diesel passenger cars, which are currently a small fraction of the US market. Stringent US diesel particulate matter emission standards represent a significant technological hurdle however.
- It is likely that the current voluntary fuel efficiency improvement measures at work in the majority of EC member states will be replaced by either a carbon dioxide emission standard (based on a life-cycle analysis) or a mandatory system of fuel efficiency standards (eg miles/gallon). Without mandatory standards, it is unlikely that adequate technology will be developed to control both conventional pollutants (especially NO_x) and to improve efficiency. Efforts in the UK to limit demand and foster technological improvements by increasing fuel taxation will have to be continually evaluated for their effectiveness.

Stratospheric Ozone and Transport

- It is likely that as CFC use is phased out and the pressure to phase-out substitutes known to contribute to ozone depletion becomes greater, the residual amount of CFCs in various end-uses will become targets for recovery and recycling programmes.
- Given that the UK passenger car fleet is relatively un-air conditioned, it is unlikely that a CFC recovery and recycle programme will be developed in the UK.
- Policies covering the servicing of existing CFC-based equipment in the refrigerated freight fleet should be anticipated, however.
- Research is required to identify the need for and potential size of CFC and halon 'banks' in the freight fleet. The implementation of

programs to establish, service and operate these banks through the development of CFC recovery and recycle programmes must also be investigated.

9.4.3 Future Tailpipe Emission Control Policies

The '94 US and '96 EC Standards

With the proposed 1996 EC standards of 0.5 g/km HC+NO_x, 2.2g/km CO, and 0.08 g/km PM, the EC has implemented tailpipe emission standards on a par with the next phase of the US 1994 standards. The major differences that exist between the two sets of standards are:

- The US will maintain a two year lead in implementing the next phase of emission standards.
- The EC maintains a combined HC + NO_x standard. These standards continue to be separate in the US.
- The EC has a slightly more stringent CO emission standard than the US – 2.2 g/km compared to 3.06 g/km.
- Particulate matter standards for diesel vehicles in the US and California remain more stringent than those for the EC – 0.06 g/km US to 0.08 EC.

The '94 California and '96 EC Standards

When compared to the 1994 California standards, the 1996 proposed EC standards can be put into proper technological perspective. The '94 California HC and NO_x standards are significantly more stringent than the 1996 EC standards. Perhaps more significant than the differences in the numerical emission standards are other refinements in the California programme of control.

- The California transitional low emission standards (TLEV) for an arithmetically combined HC+NO_x standard is 0.38 g/km compared to 0.50 g/km for the EC.
- This difference is even greater when comparing the '94 California low emission vehicle standard (LEV) to the proposed EC standard – 0.25 g/km to 0.50 g/km.
- The California standards are likely to be adopted by other US states, bringing the US market share for the California-standard cars to well over 50% of the market.
- The California programme introduces an important 'fleet averaging' method of compliance with the HC standard (non-methane organic

gases). 80% of the fleet must meet the numerical standard on an individual vehicle basis, but the rest of the fleet can comply by having their emissions averaged into the overall fleet. The fleet's average standard is then reduced incrementally on an annual basis. This will allow much more flexibility on the part of manufacturers in terms of compliance and the introduction of new models.

- The California and US programmes have moved beyond an unspecified HC standard to specify either non-methane hydrocarbons (US) or non methane organic gases (Calf.). Both of these more specific chemical representations of hydrocarbon standards are meant to focus control efforts on the reactive portion of the hydrocarbon emission stream.

- The California standards refine this principal one step further by allowing for an adjusted emission rate dependent on the reactivity of the non methane organics emitted. For example, if vehicles are fuelled with M85 (85% methanol), compressed natural gas or liquefied petroleum gas, they will have a reduced emission limit according to the reactivity of the emissions produced from vehicles using these fuels.

- Both of these methods reward emission control strategies which lower the most reactive component of vehicle exhaust pollution. They also create an incentive for introducing fuels that will reduce ozone formation.

- Another major development in the California system is the development of an emission credit scheme similar to that implemented in US fuel economy standards. For manufacturers who have a sales-weighted emission average lower than the allowable fleet average, emission credits will be granted that can be used against future years. These credits will be marketable between manufacturers to some degree. This scheme thus rewards early compliance.

The Next Generation of Standards

Both the current California law and the new US Clean Air Act achieved an important milestone in terms of emission reduction planning. Not only did they succeed in updating existing tailpipe standards, but they were successful in establishing the next generation of standards beyond the new updates.

The Clean Air Act includes so-called 'Tier II' standards scheduled to be implemented in 2004 (if necessary) and the California law will continue the TLEV and LEV programme, beginning in 1994 by enforcing more stringent standards in 1999 (ultra-low emission vehicles (ULEV) and zero emission vehicles (ZEV).

For the first time, auto manufacturers have clearly defined numerical emission targets set set in advance to allow for necessary design time. In the case of the federal standards, a full 15 year lead time is provided for the 2004 standards and, even then, they will be phased in over several years.

Other notable developments in the 2004 US standards and the 1999 California standards include:

US 2004 standards

- The US '04 standards represent a further 50% reduction in the HC, NO_x, and CO passenger car standards from the 1994 emission standards. Yet as drastic as these reductions appear, they will be technically overshadowed by the requirements of the California standards which will take on an increasingly large share of the US market. For example, the US 2004 HC standard will have been in place in California cars for ten years, the NO_x standard for eight, and the CO standard for four years.

California 1998 and beyond

- These standards will set the pace for passenger car tailpipe emission controls.
- The most notable development is the requirement in California (and those states that adopt the California programme) for 'zero emitting vehicles'. By 1998, 2% of manufacturers sales must comply with zero emissions of HC, NO_x, CO, and PM. This percentage is increased to 10% of new cars by 2003.

Future Directions for the UK and the EC

To the degree that developments in the US and California illustrate the technical potential of passenger car emission reduction and to the degree that the environmental situation requires similar actions in the EC and the UK, the following future directions can be gleaned from US developments.

Stringency

- It is clear that the 1996 EC standards should be viewed as interim measures and that more stringent targets will be developed shortly.

Given the trend towards universal standard stringency, the developments in the US should set the target for the next and subsequent rounds of EC standards.

- It is likely that EC and UK manufacturers will not be granted the lead time that the new US developments have granted US manufacturers and importers.
- A 'zero emitting vehicle' programme, either implemented as part of urban pollution control strategies or as part of fleet programmes, will most likely be introduced in some EC country within the near future to the degree that EC law allows. The Commission and member states should fund suitable pilot projects in this area and should seek to encourage participation in the new US market.

Emission Averaging and Standard Phase-in

- It is likely that as standards become more stringent, they will be phased-in over time in terms of the proportion of the fleet that has to comply. It is also likely that some sort of emission fleet-weighted average approach, similar to that implemented in California, will be developed in the EC.

Marketable Emission Credits

- It is unlikely that a system of trading emission reduction credits between mobile and stationary sources or between manufacturers of mobile sources will be developed in the UK or EC without the development of more localized air pollution control targets and plans. The latter development is a necessary one, while the former is a way to optimise control strategies.
- Such integration could take place in VOC control strategies, where mobile and stationary source emissions are already being considered together.

Focussed Photochemical Oxidant Policy

- Further HC and NO_x emission control policies, which are implemented in part to reduce photochemical smog, are likely to be based on a refined acknowledgement of the photochemical oxidation creation potential of various hydrocarbons. It is probable that the next round of EC standards could move to non methane hydrocarbons or organic gases, rather than total hydrocarbon emissions.
- The recognition of the contribution of various types of fuels towards certain environmental problems will improve dramatically over the

next few years. This knowledge will affect transport environmental policy. The CO_2 generating potential of various fuels is one such example. The investigation of alternative fuels for their impact on HC, NO_x and CO emissions is another.

- It is likely that HC and NO_x emission standards in the EC will be separate in the future as this allows for better optimization of emission control strategies in recognition of varied environmental goals.

9.4.4 Vehicle Emission Controls Beyond the 'Tailpipe'

Inspection and Maintenance (I/M) Programmes

- Inspection and maintenance motor vehicle control programmes are very cost-effective relative to other incremental emission reduction policies. As road transport grows in the UK and as more of the fleet becomes controlled by catalysts, the benefit of insuring proper maintenance and operation of the emission control system will increase.
- The addition of a loaded test cycle to the current idle test will bring the I/M programme more in line with existing test cycle procedures which have recently been changed to acknowledge actual driving conditions.
- Centralized I/M test centres could be established on a pilot project basis in order to assess the difference in emission control benefits between government operated central test facilities and private operated decentralized facilities.
- The initial test for new cars, could be moved up from 3 years to 2 or even one year in order to locate catalyst malfunctions at an earlier date.

Evaporative Emission Controls

- It is likely that the EC will adapt a 'running loss' component into its existing evaporative emissions test procedure.

Cold Temperature Carbon Monoxide Control

- CO levels in the UK are likely to continue at levels higher than the EC and WHO directives and guidance levels at colder times of the year. A cold temperature CO standard could improve the overall ambient CO problem.

- If a cold temperature CO standard is adopted (and perhaps even if it is not), lower CO motor vehicle emissions will most likely be achieved through a mandated minimum percentage of oxygenate added to petrol (between 2–4% weight by volume). Preliminary results from the US indicate the enormous potential of oxygenated fuels for reduction of vehicle-related CO emissions.

- Such fuels-based emission control policies may be good to investigate for particular climates and regions that are experiencing air quality problems which other areas are not. The ability of fuels-based programmes to address niche markets vehicle-based programmes cannot should be investigated further.

- The addition of an ambient cold-temperature test condition to the vehicle performance test cycle would ensure that vehicles are designed with all the appropriate operating conditions in mind.

9.4.5 Changes in Vehicle Fuel Policies

Lead in Petrol

- Regular grade leaded petrol will eventually be decreased in market share to the point where a ban for road transport use (excluding agricultural machinery) will take place. The increased use of auto catalytic converters will be the main incentive.

- It is doubtful that allowable lead limits in leaded petrol will be reduced from the current 0.15 - 0.40 g/litre. The move of countries to prohibit the sale of leaded petrol altogether will make such a move unnecessary.

Benzene

- As the health effects of mobile source air toxics become a bigger part of the transport environment agenda in the EC and the UK, benzene content in gasoline will be a primary target for further regulation.

- It is unlikely that ambient benzene standards will be set, but guide values similar to those used for ozone and lead could be enacted at the EC level. It is more likely the fuel content restrictions will be tightened. For the moment, a 1% maximum content by weight should be considered the reasonable near-term regulatory target.

- The implications on fuel volatility and other fuel quality parameters, as well as the implication on refinery production of a lower benzene content, should be investigated.

Reformulated 'Cleaner' Gasoline

- The ability of 'cleaner' petrol to assist in the reduction of NO_x emissions should be closely monitored. If this aspect is successful, then UK fuel reformulation could develop along these lines.
- As knowledge of the ambient and urban levels of hazardous air pollutants increases in the UK, and as the health effects of these pollutants become more of a concern for UK policy makers, the attractiveness of reformulating fuels from an air toxic perspective becomes greatly enhanced.
- Volatility control for environmental reasons will take greater prominence in the EC and will eventually result in significantly reduced petrol volatility. Quantification and regulation of evaporative emissions from vehicles will also act to create pressure from the auto manufacturers for lower volatility fuels.

Diesel Fuel Quality

- Based on developments elsewhere, a move to 0.05% maximum diesel fuel sulphur content in the UK is most likely.
- Diesel fuel content regulations are becoming increasingly specific with limits on aromatic content and cetane prevalent.
- As diesel fuel quality becomes more specific, pressure will grow on governments to move to diesel fuel 'performance' standards, not unlike the reformulated gasoline program in the US, in order to maintain refinery flexibility and to ensure sufficient capacity for the growing diesel market.

Oxygenated Fuels

- The use of increased amounts of oxygenate in motor vehicle fuels in the UK will most likely mirror the development of oxygenated fuels policies in the EC.
- If exceedence of carbon monoxide ambient standards and guidelines continues or increases, it can be expected that mandated levels of oxygenates in petrol will be seriously looked at as a regulatory solution.

9.4.6 New Types of Policy Responses

Addressing the 'old polluting car' problem

As the incremental cost of reducing vehicle emissions from new cars continues to increase and the relative merit of such improvements decreases, governments are likely to look toward methods to address the 'old polluting car'. When they do, there are two methods being tried in various locations that will have an impact on the UK.

Accelerated Vehicle 'Scrappage'

- Accelerated scrapping, or getting old cars off the road prior to their anticipated date of being 'junked', is one such method. Accelerated scrappage programmes offer a unique market-based solution to localized air pollution programmes. These scrappage programmes are being implemented in areas that have specified emission reduction targets for their particular areas.
- In the UK, where local regions do not have localized ambient air quality goals, this motive force is missing. Although nation-wide air quality goals in the UK can provide the framework with which to evaluate such programs, a formal recognition of the relationship between local air quality improvements and national goals is still needed. Accelerated scrappage programmes could have an impact on the local economy and potentially the environment. Implementing a pilot programme would create a demand for much needed vehicle fleet and driving condition information.

Retrofit Programmes

- Retrofit programmes can also help in crafting local solutions to local problems. However, unlike scrappage programmes, the benefits of vehicle retrofit are likely to be greater and the programme easier to promote and analyze.
- Bus and cab fleets are the most likely candidates for retrofit programmes. Although the local air quality improvement imperative that exists in the US is missing in the UK, where improvement of urban air quality is an objective of local government, retrofit programmes could be quite attractive.
- Further bus privatisation around the country could, as a condition for sale and route access, require improved environmental performance of the fleets, retrofit could play an important role.

Off-Road Vehicle Emissions

- The prevalence of off-road vehicle emissions (eg agricultural tractors) in the UK emissions inventory will increase over time. Regulations to address these sources should be welcomed and are being developed in the EC and UNECE.

- As attention turns toward off-road and stationary sources, it is likely that more encompassing air pollution control strategies and systems will be developed that acknowledge the contribution of all sources of air pollution.

Mobile and Stationary Source Emission Trading

- Emission credit and trading schemes will elicit investigation when optimization of pollution control is required. The potential economic benefit of such programs is great. It is estimated that the new acid rain emission trading programme in the US will save approximately $1 billion or 25% of the cost of a non-trading system. Perhaps the greatest benefit of any emissions trading scheme is the incentive it gives for cost-effective technology development.

Vehicle Emissions and Vehicle Speed

- Reducing motorway vehicle speed meets multiple policy objectives. It reduces the number of motorway fatalities as experience in the US has proven. It saves energy and thus reduces CO_2 emissions. It also reduces conventional pollution emissions such as HC and NO_x. Under most conditions, it also reduces road-tyre interface noise.
- Efforts to enforce existing speed limits are being increased. A more effective policy would be to either lower UK speed limits, require the installation of speed limiters on passenger cars, or both. Speed limiters would require EC legislation and are unlikely to be considered due to likely consumer unacceptability.